Mathematical Engineering

Today, the development of high-tech systems is unthinkable without mathematical modeling and analysis of system behavior. As such, many fields in the modern engineering sciences (e.g. control engineering, communications engineering, mechanical engineering, and robotics) call for sophisticated mathematical methods in order to solve the tasks at hand.

The series Mathematical Engineering presents new or heretofore little-known methods to support engineers in finding suitable answers to their questions, presenting those methods in such manner as to make them ideally comprehensible and applicable in practice.

Therefore, the primary focus is—without neglecting mathematical accuracy—on comprehensibility and real-world applicability.

To submit a proposal or request further information, please use the PDF Proposal Form or contact directly: Dr. Thomas Ditzinger (thomas.ditzinger@springer.com)

Indexed by SCOPUS, zbMATH, SCImago.

Xiu Ji

Spin, Speed and Strategy

A Mathematical Guide to Table Tennis

Springer

Xiu Ji
Department of Engineering
Manchester Metropolitan University
Manchester, UK

ISSN 2192-4732 ISSN 2192-4740 (electronic)
Mathematical Engineering
ISBN 978-3-032-25517-4 ISBN 978-3-032-25518-1 (eBook)
https://doi.org/10.1007/978-3-032-25518-1

This Springer imprint is published by the registered company Springer Nature Switzerland AG
The registered company address is: Gewerbestrasse 11, 6330 Cham, Switzerland

Dedicated to the players of the Manchester, Wilmslow and Trafford leagues, who provided the laboratory for these ideas.

Preface

In the millisecond-scale environment of a table tennis rally, the observer witnesses a paradox: a sport that is intuitively accessible but mathematically formidable. The outcome of elite play hinges on nuances often invisible to the naked eye, minuscule variations in the coefficient of friction, subtle shifts in angular momentum and the deliberate modulation of temporal rhythm. For generations, coaches and athletes have approached this complexity through the subjective realm of 'feel'. This monograph seeks to transition that wisdom into the domain of formal engineering.

Origins

The core inspiration for this book took shape during a period of regular lunch-break games at my university's Students' Union. I recall a particular afternoon when, as I walked through the university park towards the hall, several students stopped me to explain that it was closed for an event. Their immediate recognition and certainty that I was heading for the tables made me realise how closely the sport had become part of my life. In that moment, I first considered applying physics to illuminate table tennis for my students, from the angles of mechanical forces, motion and trajectories.

It soon became apparent, however, that physics alone was insufficient for securing victory in the competitive 'winner-stays-on' environment of a crowded hall. A higher-level tactical and conceptual framework, one connecting real-time battle and cognition, was required to elucidate the rationale behind the interactions observed during play. This book grew out of that realisation, developed over several years through scholarly inquiry, practical exploration, reflection and iterative revision.

The Objective of this Work

This book introduces the Table Tennis Notation System (TTNS): a formal symbolic language designed to render the hidden dimensions of the sport measurable and computationally analysable. A single example captures the spirit of this: the symbol 100 encodes a slow topspin stroke, a concrete handle on something that would otherwise live only in a player's muscle memory. TTNS functions as a bridge, connecting the phenomenological experience of play with the rigorous principles of classical mechanics and dynamical systems theory.

Central to this bridge is the use of a small number of primitives to construct more complex attributes. A concise set of spin symbols, σ (topspin), $\reflectbox{$\partial$}$ (backspin) and θ (sidespin), is introduced as formal operators. Their visual clarity and compositional structure allow compound spin states such as $\sigma\theta$ (dominant topspin with sidespin) to be expressed unambiguously, where natural language would require cumbersome description.

The guiding question of the framework is a practical one: can a table tennis rally be described with sufficient clarity to admit mathematical or computational analysis? A productive answer lies in the recursive event chain $\mathrm{I_n} \rightarrow \mathrm{F_1} \rightarrow \mathrm{TC} \rightarrow \mathrm{F_2} \rightarrow \mathrm{I_{n+1}}$, in which each interaction both concludes and initiates the next. Once match play is encoded in this discrete sequence, the apparent complexity becomes tractable. The multitude of serves, receives, exchanges and tactical patterns is then organised into a structured form amenable to rigorous analysis.

Building on this representation, the ε-strength factor provides a calibrated measure of the intensity of a discretised attribute (e.g., force, velocity, or tactical tendency), whilst the δ-operator captures variation across placement, timing, speed, spin and rhythm.

Although the framework developed here is compatible with computational analysis and artificial intelligence, its purpose is not to replace human decision-making but to equip players with fundamental knowledge that can be internalised and applied under the pressure of real play. The book follows a single thread: physical interaction produces stroke quality; stroke quality shapes the tactical state; the tactical state evolves through uncertainty; uncertainty can be represented symbolically; and symbolic representation enables analysis, training and AI integration.

How to Read This Book

The text is structured to guide the reader from the physics of ball–bat interaction to the logic of tactical play. Chapters 2 and 3 establish the mechanical foundation: equipment properties (hardness, stiffness, friction), aerodynamic forces, and collision models, including spring–mass–damper and dual-force representations. These are applied to impact dynamics, ball flight and bat restitution. Chapter 4 translates that foundation into tactical structure, linking strokes, serves and rally patterns to geometry, energy, timing and decision-making.

From this base, Chaps. 5 to 10 introduces the core of TTNS, a framework for discretising strokes, serves and tactical intent, encoding shot intention through a concise lexicon in which, for example, 100 denotes a slow topspin stroke and ə the backspin axis. Subsequent chapters develop the tactical dialectic, using the δ-variation measure to quantify the accumulation of disruption within a rally and the resulting cognitive and perceptual information load (entropy E_p) imposed on the opponent during a rally.

The chapters are best read in sequence. Core concepts recur throughout, each time at a higher level of formalisation: principles introduced at the level of physical interaction are progressively reinterpreted through action construction, tactical structure and match dynamics.

A Reflection on Formalism

One might ask if such an engineering approach is necessary for a sport so naturally played. As the research for this book progressed, however, it became clear that the mathematical formulae are not imposed upon the sport; they are inherent in its fabric. The ball moves not by 'feel' but by the exact geometry of force. By introducing engineering logic into this artistry, our understanding may be enriched.

This work is written for readers who wish to understand table tennis as a technical system, whether they come from sport, engineering, coaching, or analytics. Its scope may appear broad, but its inquiry is singular: what does it actually take to win? The game punishes ignorance at a speed that leaves no time for conscious correction. Understanding is perhaps the only sustainable advantage in a sport where action so often precedes conscious thought. It is my hope that this formal language provides a new perspective on the sport, deepening both the systemic understanding and the enduring appreciation of table tennis.

Manchester, UK
April 2026

Xiu Ji

Acknowledgements I would first like to acknowledge the wider table tennis community: those with whom I played in public parks and community halls and the many opponents encountered in the Manchester leagues. Their diverse styles, idiosyncratic habits and tactical adaptations provided the empirical grounding for this analysis. Table tennis is an interactive system; many of the ideas formalised herein were developed through the sustained observation of how players respond under the combined pressures of competition, timing and mechanical demand.

I am grateful to the executive editor and the Mechanical Engineering series editors for their prompt response to the original book proposal. Their description of the project as "refreshing", and their encouragement to expand the initial short-book proposal into a fuller monograph, helped me recognise the appropriate scale of the work. At a point when I was still trying to compress the material into journal papers or a shorter volume, their confidence in the project provided timely reassurance and momentum.

I am particularly grateful to my family and friends. Initially amused by my preoccupation, they later became the sounding board for my informal 'lectures' as my thinking progressed from basic physical principles to the abstraction of game elements through symbols and notation. Their inquisitive challenges helped these concepts mature into the structured narrative now presented to the reader.

Finally, I am deeply indebted to those close to me who tolerated the protracted and uneven progress of this project. Their patience during repeated returns to unfinished chapters made this work possible.

Competing Interests The author has no competing interests to declare that are relevant to the content of this manuscript.

List of Symbols

The notation used in this monograph integrates physical variables with the symbolic constructs of the Table Tennis Notation System (TTNS). Latin and Greek symbols denote mechanical and kinematic quantities, whilst structured symbols such as (J, S) and (E, G) represent higher-level technical and tactical states.

Several symbols are used in a *context-dependent manner*, carrying distinct meanings across mechanical, geometric and TTNS domains. This dual usage is intentional and reflects the layered structure of the framework. In all cases, meaning is unambiguously determined by context, and each interpretation is defined at first occurrence.

In particular:

- e denotes elastic (Young's) modulus, whilst E denotes energy. Symbols are case-sensitive.
- r denotes a mechanical lever arm or a temporal rhythm variable.
- W denotes blade width or gravitational force (weight).
- h denotes blade thickness or ball wall thickness.

F_h	Hitting force, acting through the centre of the ball, normal to its surface
F_b	Brushing force, tangential to the ball surface
v	Translational velocity of the ball (speed)
ω	Rotational velocity of the ball (spin)
U	Ball kinematic state $U = (v, \omega)$
t_d	Dwell time, impact interval, contact time
H	Hardness
F	Force, or total force of ball–bat impact $F = F_h + F_b$ (context-dependent)
A	Area (typically the cross-sectional area of the ball)
$\varDelta$	Elastic deflection
k	Stiffness
ζ	Damping ratio (dimensionless)
k_I	Stiffness of the bat
c_I	Damping coefficient of the bat (viscous damping)

m_1	Mass of the bat or blade (context-dependent)
ζ_1	Damping ratio (dimensionless) of the bat
k_2	Stiffness of the ball
c_2	Damping coefficient of the ball (viscous damping)
m_2	Mass of the ball
ζ_2	Damping ratio (dimensionless) of the ball
k_e	Effective stiffness of ball–bat coupled spring–mass–damper system (SMD)
R	Ball radius, $R = 20$ mm
e	Young's modulus (elastic modulus)
L	Blade head length
L_1	Blade handle length
W	Blade width or gravitational force (context-dependent)
h	Blade thickness or ball wall thickness (context-dependent)
h_0	Blade core ply thickness
h_1	Blade outer ply thickness (Ply 1)
h_{-1}	Blade opposite outer ply thickness (Ply 1)
I_A	Second moment of area
d	Distance from ply centroid to neutral axis
υ	Poisson's ratio
λ	Modal constant
D	Flexural rigidity
ρ	Material density
f_1	Fundamental natural frequency (first mode)
r	Lever arm (mechanics) or rhythm (context-dependent)
T_{q1}	Torque generated by the brushing force, $T_{q1} = R \times F_b$
T_{q2}	Torque generated by deformation during ball–bat impact, $T_{q2} = r \times F_h$
J_h	Normal impulse
J_b	Tangential impulse
J	Total impulse, $J = J_h + J_b$ or impulse state $J = (J_h, J_b)$
I_m	Mass moment of inertia of the ball
β	Bat angle
F_f	Frictional force
F_n	Normal force
μ	Coefficient of friction
μ_0	Static coefficient of friction
μ_k	Kinetic coefficient of friction
μ_e	Effective coefficient of friction
CoR	Coefficient of restitution
R_r	Reactive Return
A_r	Active Return
α	Trajectory angle or ball direction
g	Gravitational acceleration, $g = 9.81$ m/s^2
σ	Topspin
∂	Backspin

θ	Sidespin
χ	No-spin
S	Spin-to-speed ratio $S = \omega \cdot R/v$
(J, S)	Stroke quality (mechanical execution)
$\varphi\ (J, S)$	Stroke quality
$\varphi_{opp}(J, S)$	Technical pressure for the opponent
E	Ball's total kinetic energy, $E = E_v + E_\omega$ or energy state $E\ (v, \omega)$
E_v	Translational kinetic energy
E_ω	Rotational kinetic energy
P	Power transferred to the ball during impact
I_1	Player's (first) contact in an exchange
F_1	First flight phase
TC	Table Contact (bounce)
F_2	Second flight phase (after bounce)
I_2	Receiver's (first) contact
F_d	Aerodynamic drag
C_D	Drag coefficient
K_d	Drag parameter, $K_d = \rho \cdot C_D \cdot A/2$
F_m	Magnus force
C_L	Lift coefficient
C_m	Magnus coefficient, $C_m = C_L/S$
K_m	Magnus parameter, $K_m = \rho \cdot C_L \cdot A/2$
Re	Reynolds number
ϑ	Kinematic viscosity of air ($\approx 1.46 \times 10^{-5}$ m^2/s)
N_H	Net height, $N_H = 15.25$ cm
T_L	Table length, $T_L = 2.74$ m
T_w	Table width, $T_w = 1.52$ m
T_H	Table height, $T_H = 76$ cm
τ	Interception timing ($\tau_{rising}, \tau_{peak}, \tau_{falling}$)
t_w	Reaction Window, Response Window, or Reaction Interval
(α, z)	Geometric state; $G = (\alpha, z)$
(E, G)	Tactical state, $T = (E, G)$
T_R	Targeted tactical state, $T_R = (E_R, G_R)$
E_p	Entropy (information measure)
p1	Order-imposing playing regime
p0	Chaos-imposing playing regime
1	Digit 1: thesis (e.g. Hit, Speed, Tactic1)
0	Digit 0: antithesis (e.g. Brush, Spin, Tactic0)
ε	E15 strength factor
ε_1	Strength of speed (translational component), or ε_v
ε_0	Strength of spin (rotational component), or ε_ω
ε_H	Strength of hitting force (normal component)
ε_B	Strength of brushing force (tangential component)
T_1	Tactic1 (Winner strategy)
T_0	Tactic0 (Error-forcing strategy)

ε_{T1}	Strength of Tactic1
ε_{T0}	Strength of Tactic0
δ	Variation of play
δ_z	Variation in placement
δ_τ	Variation in interception timing
δ_1	Variation in speed, or δ_v
δ_0	Variation in spin, or δ_ω
δ_r	Variation in rhythm

Contents

About the Author

Xiu Ji is Senior Lecturer in Engineering at Manchester Metropolitan University. She is Chartered Engineer (CEng), Member of the Institution of Engineering and Technology (MIET), and Senior Fellow of the Higher Education Academy (SFHEA).

Holding a PhD in Artificial Intelligence and drawing upon extensive industry experience at Siemens Automation and Drives, she has taught Mathematics, Physics, Control Engineering and Systems Engineering at Manchester Metropolitan University for over a decade. Her research and teaching focus on new perspectives on classical physics and mathematical modelling. Outside academia, she is Accomplished Amateur Table Tennis Player with a profound interest in history and philosophy.

List of Figures

List of Tables

Chapter 1
Introduction

Table tennis is deceptively simple. A net, a lightweight ball and two players suggest, at first glance, little more than a casual pastime. Behind this minimalist setup lies a sport defined by extreme velocities, dense interactions and a level of technical subtlety rarely apparent to the untrained eye. It is a game that almost anyone can learn within minutes, yet one that denies consummate command even after decades of practice.

This tension between accessibility and complexity is at the heart of the game's enduring appeal. Players at every level encounter non-linear phenomena: a ball whose spin is suddenly unreadable, a rally that alters its character in a single shot, or a rhythm that shifts imperceptibly before becoming decisively uncomfortable. Progress is tangible, but neither linear nor guaranteed; mastery appears simultaneously attainable and perpetually incomplete.

Such paradoxes have long attracted writers, coaches and scientists alike. As coach and author **Matt Hetherington** observes, "Table tennis is like an atom: to the ignorant it is merely microscopic and insignificant but to the dedicated it is intricate in design and the building block to everything we know". The Nobel laureate **Werner Heisenberg**, himself a keen player, was drawn to a closely related principle: how minute variations in spin and angle can generate disproportionately large, non-linear effects in a ball's trajectory – an evolution of a rally that often defies intuition (Schneider et al. 2022). Beyond the physics of the table, early advocates such as **Ivor Montagu** emphasised the sport's unusual combination of physical agility and mental clarity (Griffin 2014). Others have drawn parallels to chess, wherein each move constrains the manifold of future options (Hodges 2013) or to music, where expertise is cultivated through deliberate, structured practice over time (Ericsson et al. 1993).

The fascination arises because infinitesimal physical differences in timing, angle, speed and spin can produce divergent outcomes, making minute variations paramount.

X. Ji, *Spin, Speed and Strategy*, Mathematical Engineering,
https://doi.org/10.1007/978-3-032-25518-1_1

This work develops an analytical framework that links the mechanics of ball–bat interaction to the informational structure of rallies. Its purpose is to make visible what has until now remained implicit: that winning and losing in table tennis are not random events but the measurable consequences of physical laws, tactical logic and structured decision-making. For the curious player, it offers a way to understand why some losses feel inexplicable and some victories come as a surprise. For the coach, it provides a disciplined vocabulary that complements and sharpens experiential knowledge. For the researcher and sports scientist, it offers a formal structure against which alternative methods and models can be compared.

The Fundamental Mechanics of Interaction

At the core of the game lie two fundamental physical actions: the hitting force (F_h) and brushing force (F_b).

- **The hitting force** (F_h): a direct impact applies a predominantly normal force, imparting translational velocity (v) to the ball.
- **The brushing force** (F_b): a brush applies a predominantly tangential force, generating rotational velocity (spin, ω).

They are treated as dialectical opposites from which the entire game is constructed, and the tension between them serves as a generative principle throughout the book. In discrete form, this duality admits a minimal encoding: $F_h \equiv 1$, $F_b \equiv 0$.

Their combination gives rise to the full technical vocabulary of strokes, serves and receives; from their opposition gives structure to the tactical dialectic of order and chaos that defines competitive play at every level.

The Table Tennis Notation System (TTNS)

The Table Tennis Notation System (TTNS) employs binary distinctions and structured notation to delineate strokes, serves, rallies and tactical sequences.

At its foundation is the minimal encoding introduced earlier: the hitting force (F_h) as 1 and the brushing force (F_b) as 0. From this binary opposition, more complex stroke compositions and tactical patterns can be constructed and analysed within a unified formal system.

As algebraic chess notation allows games to be reconstructed and audited, and musical notation allows performances to be preserved, transmitted, taught and studied, the TTNS provides a concise language for expressing the latent logic of the sport. Its core lexicon encodes shot intention concisely: 100 (Hit–Brush–Brush) denotes a slow topspin stroke; ə encodes the backspin axis; stroke magnitude is parameterised by a dual-component ε-strength factor, ε_H and ε_B, each expressed on a 1–15 scale to enable consistent comparison across strokes. Table tennis has lacked such a language. Without it, analysis has depended on description and coaching on intuition. The TTNS does not aim to replace either. It aims to make both more powerful by rendering structure visible and linking physical action and tactical consequence within a single analytical framework.

In doing so, it establishes an interpretable representation of play that is naturally compatible with computational analysis, providing a foundation upon which data-driven models and artificial intelligence systems may be constructed.

Intense competition in table tennis invites comparison with Game Theory, in which decision-making under opposition is formulated as optimisation over predefined strategies and payoff structures. The present framework is different, drawing instead on three theoretical traditions: control theory, dynamical systems and information theory.

From control theory comes the language of state regulation, stability and drift. From dynamical systems comes the treatment of play as a physically evolving system in which strokes are constructed through impulse and spin, carrying inherent sensitivity and risk of breakdown. From information theory comes the concept of entropy as a measure of uncertainty, one that is actively modulated across the course of rally through the δ-operator, which quantifies variation across five dimensions: placement, timing, speed, spin and rhythm.

These three lenses reframe what strategy means in table tennis. It is not an optimisation over fixed payoffs but an emergent organisation of actions, shaped in real time realised through execution. Performance depends on control stability and execution fidelity. And uncertainty arises from the mechanics of physical interaction itself.

The Journey Ahead

The chapters that follow trace a progression from physical principle to systemic interpretation. They are structured as a sequence and are best read in order.

Physics to Code

Chapters 2 and 3 constitute the analytical core of the book. They present a quantitative examination of equipment, contact mechanics and ball flight through the lens of classical mechanics. Phenomena such as the 'catapult effect' are modelled explicitly, whilst concepts often treated informally, such as the 'ghost ball', are defined in rigorous physical terms.

Chapter 4 then lifts this physical foundation into the strategic domain. It examines how strokes, serves and tactical patterns are shaped by mechanical conditions and organised through the coordinated control of geometry, energy and timing. Impacts are no longer treated as isolated mechanical events, but as components of rally construction and are combined, adapted and sequenced within play. Chapter 4 therefore establishes the operative structure upon which the subsequent formalisation of TTNS is built.

Across these sections, empirical findings from the scientific literature are introduced as representative ranges, material properties and measured coefficients. This provides realistic physical bounds for competitive play and anchor qualitative concepts such as hardness, stiffness, friction and restitution to experimentally observed scales. The approach reflects engineering practice: variables are interpreted through magnitude, variability and domain of validity.

Code to System

Table tennis is shaped by persistent oppositions: speed and spin, hitting and brushing, stability and disruption, instinct and intention. Chapter 5 formalises these as dialectical tensions. The term is used in a practical rather than doctrinal sense. It refers to structured oppositions between mechanical and tactical tendencies whose resolution gives rise to usable technique, adaptable style and more deliberate play.

Chapters 6 to 10 introduce a formal coding framework for strokes, serves, tactics and rhythmic patterns, integrated progressively into a unified model of table tennis as a structured dynamical system. Strokes are not classified as isolated techniques, but as related actions distinguished by impulse profile, spin ratio and tactical intent. Higher-order constructs, including stability, variation, and entropy-based descriptions of tactical interaction, are thereby accommodated within a single formal system capable of supporting systematic analysis, computational modelling, simulation and prediction, and interpretable artificial intelligence grounded in physical causality. In practice, this supports match preparation and rally decomposition for analysts, reproducible objective feedback for coaches, structured self-analysis for players, and a physically grounded discrete grammar of play for AI systems engaged in sequence modelling and prediction.

This formulation provides a compact language through which players, coaches and analysts can internalise the structure of table tennis, precise enough to survive the time-compressed conditions of real play, yet tractable enough for computational representation. Its value lies in what it makes thinkable, not only in what it makes measurable. Match analysis is one application of that language, not its destination.

Chapters 11 and 12 then return the framework to the full game. Chapter 11 applies layered decomposition to match analysis, showing how physical, technical and tactical states interact across real sequences of play. Chapter 12 interprets table tennis as a systemic control problem, drawing together structural tensions, TTNS-based regulation, adaptability and tolerance of uncertainty, and the deeper fulfilment that comes from understanding the game as an integrated whole.

The book develops a technical language for table tennis: from contact mechanics and flight dynamics, through stroke construction and tactical interaction, to symbolic representation and player development. Its objective is to bridge intuition and analysis, providing analytical clarity as a foundation for deeper understanding and more sustained enjoyment.

References

Ericsson KA, Krampe RT, Tesch-Römer C (1993) The role of deliberate practice in the acquisition of expert performance. Psychol Rev 100(3):363–406. https://doi.org/10.1037/0033-295X.100.3.363

Griffin N (2014) Ping-pong diplomacy: Ivor Montagu and the astonishing story behind the game that changed the world. Simon & Schuster, London

Hodges L (2013) Table tennis tactics for thinkers. Lulu Press, Morrisville, NC
Schneider R et al. (2022) Table tennis and physics. In: Advances in sport technology. IntechOpen. https://doi.org/10.5772/intechopen.107026

Chapter 2
Impact Mechanics: The Origin of Speed and Spin

2.1 Background

What began as a Victorian dining-room diversion has become one of the most physically demanding and technically intricate sports in the world. As **Jenny Heaton** observes, table tennis is believed to have originated in the 1880s as an adaptation of lawn tennis playable on a dining table (Heaton 2016). From this modest origin, its material foundations have undergone continuous and consequential refinement.

Early experiments with rubber and cork spheres were superseded by the introduction of celluloid in 1900, establishing table tennis as a sport distinct from a parlour diversion. A further standardisation occurred in 2000, when the ball diameter increased from 38 to 40 mm and its mass from 2.5 to 2.7 g. This change was intended to reduce the kinetic energy carried by the ball.

The larger diameter increases the cross-sectional area and hence aerodynamic drag, while the modified mass and geometry alter the moment of inertia. Together, these effects reduce translational speed and moderately attenuate spin, resulting in longer exchanges and a subtle shift in competitive balance from speed- and spin-dominant play towards consistency, placement, control and variation. The underlying fluid-dynamic mechanism, the role of cross-sectional area in drag scaling, is examined in Sect. 3.1.2.

A further transition occurred in 2014, when plastic (Acrylonitrile Butadiene Styrene, ABS) balls replaced celluloid, primarily to mitigate flammability concerns. Although marketed as functionally equivalent, ABS balls differ measurably in elastic response and acoustic signature, subtle material variations immediately perceptible to skilled players and a remind of the sport's acute sensitivity to infinitesimal physical differences.

The most transformative mechanical innovation, however, preceded both changes. In the 1950s, the development of the 'sandwich' bat, a sponge layer inserted beneath the rubber topsheet, gave players the capacity to induce large elastic deformations during impact. This single mechanical property enabled the generation of

X. Ji, *Spin, Speed and Strategy*, Mathematical Engineering,
https://doi.org/10.1007/978-3-032-25518-1_2

strong topspin, backspin and complex hybrid spins: forms of rotational input previously unattainable. As a result, the tactical domain expanded, with compound spin becoming the defining strategic feature of modern play.

As **Guido Mina di Sospiro** observes: "Tennis remains a sport that favours the player's physical status and power. It missed out on its chance to evolve and become a more sophisticated game, unlike table tennis" (Mina di Sospiro 2015). Modern table tennis thrives on the coordinated manipulation of spin, speed, placement and temporal rhythm. The bat, though superficially simple, is a finely engineered composite whose mechanical properties critically dictate the ball's physical and tactical behaviour.

Table tennis bats are rigid, rubber-covered paddles, not stringed rackets. Their interaction with the ball is governed by solid-surface impact and viscoelastic deformation, fundamentally different from string-bed rebound. Understanding how the bat is constructed, and how its materials deform, recover and transmit force is the necessary first step. It provides the mechanical foundation for all subsequent modelling of ball–bat dynamics.

2.2 Construction of a Table Tennis Bat

A modern bat comprises three principal functional layers: a wooden *blade*, a polymeric *sponge* layer and an outer *rubber topsheet*. The thickness, elasticity and geometry of these layers collectively determine the bat's performance, commonly described in terms of *speed*, *spin* and *control*. Typical layer dimensions are illustrated in Fig. 2.1.

Additional factors, including head geometry, handle ergonomics and total mass, also influence both the subjective haptic feedback and the objective rebound dynamics. At the contact interface, these interactions are governed primarily by the bat's coefficient of restitution (*CoR*) and coefficient of friction (μ), both are defined and quantified in subsequent sections.

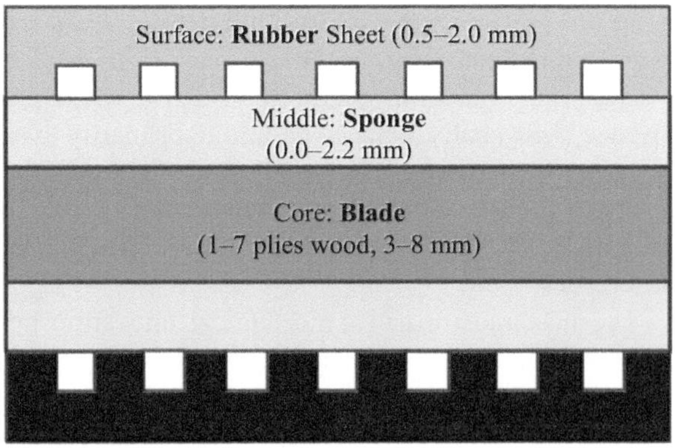

Fig. 2.1 Cross-sectional architecture of a composite table tennis bat. Schematic of a five-layer sandwich construction comprising a central wooden core (3–8 mm), flanked by elastic sponge layers (0.0–2.2 mm) and outer rubber surfaces (0.5–2.0 mm)

2.2.1 *Table Tennis Equipment: Sandwich Structure*

The Blade

The blade constitutes the rigid core and is typically constructed from 1, 3, 5 or 7 plies of wood. Thin 3-ply blades generally measure between 3.0 and 4.6 mm, whilst 5- and 7-ply variants range from 5.0 to 8.0 mm, depending on the wood species and ply orientation. Common species include balsa, ayous (obeche), limba and koto, selected for specific moduli of hardness, density and flexural stiffness.

- **Hard outer plies (e.g. limba)**: selected to facilitate a 'crisp', high-velocity rebound.
- **Compliant materials (e.g. ayous)**: utilised to increase the *dwell time* (t_d), thereby enhancing spin-generation potential through extended tangential contact.

Modern blades often incorporate composite reinforcements such as carbon fibre, arylate or aramid to augment stiffness, enlarge the effective sweet spot (the region of *uniform restitution*), and attenuate undesirable vibration modes. Per International Table Tennis Federation (**ITTF**) regulations, at least 85% of the blade's thickness must consist of natural wood. Any reinforcing layer must not exceed 7.5% of the total blade thickness or 0.35 mm, whichever is smaller. Furthermore, the total thickness of the rubber and sponge combined must not exceed 4.0 mm (ITTF (2026).

The Sponge

The sponge layer, situated between the blade and the topsheet, serves as the primary energy storage component.

- **Thick sponges (1.9–2.2 mm)**: accommodate greater elastic deformation, storing and releasing more impact energy to enhance ball velocity and spin potential.
- **Thin sponges (0.5–1.3 mm)**: reduce deformation, providing increased control and more direct haptic feedback.

At the extreme lower end of the sponge-thickness continuum, removing the sponge entirely produces a *hardbat* (Fig. 2.2c). Here, the rubber adheres directly to the blade, offering a qualitatively different response: the ball leaves the bat with reduced rebound velocity, diminished spin generation and a steeper, shorter post-impact trajectory. The hardbat rewards deliberate placement and rapid exchanges over power-based rallies.

The career of **Marty Reisman** illustrates this configuration's historical and tactical coherence. Following the rapid adoption of sponge rubber in the 1950s, Reisman remained committed not just to a hardbat blade but to the entire pre-sponge technical regime: no sponge layer, and pimples-out topsheet. His sustained success across five decades demonstrated that this older constellation was a viable mechanical and tactical system, one capable of resurgence, as younger athletes periodically rediscover niche approaches that challenge the mainstream and, in doing so, renew the game itself.

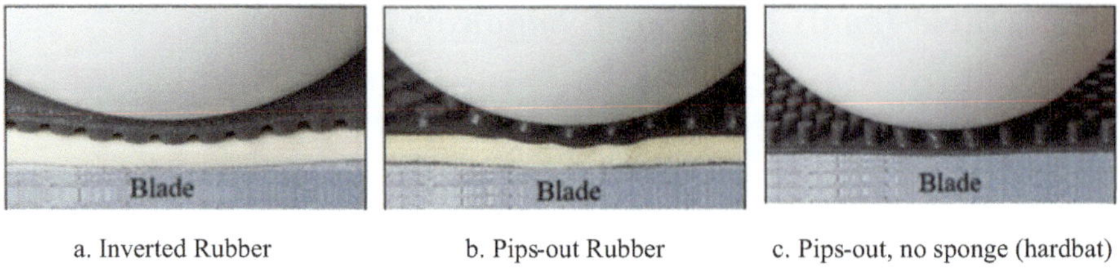

Fig. 2.2 Comparison of rubber configurations. **a** Inverted, or pips-in, rubber; **b** Short pips-out rubber with pips, with pips facing upwards above a sponge layer; **c** Hardbat configuration, with long-pips mounted directly to the blade substrate

For clarity throughout this text, *soft* may be used to denote higher material compliance (i.e. lower elastic modulus) in sponge and rubber layers, and lower *surface hardness* in blade materials. The term *flexibility* (and its inverse, *stiffness*) is reserved exclusively for the global bending behaviour of the blade as a structural system (Sect. 2.4).

The Rubber Topsheet

The topsheet (typically 0.5–2.0 mm thick) forms the primary contact surface. Rubber design manipulates polymer composition and pimple geometry (height, aspect ratio, spacing; Table 2.1) to tailor frictional traction and rebound behaviour.

- **Inverted (pips-in) rubber** features a smooth exterior with pimples-oriented inwards (Fig. 2.2a). High surface friction permits maximal spin generation and high sensitivity to incident spin.
- **Anti-spin rubber** is a low-friction variant of inverted rubber engineered to suppress tangential friction and neutralise incident *rotational velocity* (spin).
- **Short pips** utilise outward-facing pimples (Fig. 2.2b), reducing spin sensitivity and favouring a flatter, more direct playing style.

Table 2.1 Table tennis rubbers and their operational characteristics

Rubber type	Surface geometry	Rebound speed	Spin generation/ sensitivity	Primary tactical application
Inverted	Inward-facing; smooth topsheet	High	High/maximal	Versatile; proactive spin generation
Anti-spin	Inward-facing; low-traction surface	Moderate	Minimal/ neutralising	Defensive; spin neutralisation
Short pips	Outward-facing; short textured pips	Moderate	Low/reduced	Direct hitting; blocking
Long pips	Outward-facing; high-aspect-ratio pips	Low	Minimal/ disruptive	Defensive; tempo disruption

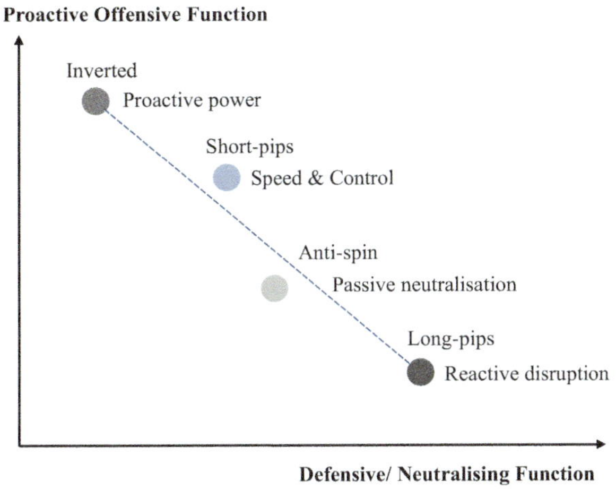

Fig. 2.3 Conceptual positioning of major rubber types by tactical function. The horizontal axis indicates disruptive or neutralising tendency, whilst the vertical axis indicates capacity for proactive play. The diagram is schematic rather than exhaustive

- **Long pips** employ high-aspect-ratio pimples (Fig. 2.2c) that bend and buckle sharply upon impact, producing erratic trajectories suited primarily for tactical disruption.

Figure 2.3 illustrates the fundamental engineering trade-off in table tennis rubber selection, mapping the continuous transition from inverted surfaces to long pips. With inverted rubber, the player dictates the pace and spin of the ball; with long pips, the player leverages the opponent's incoming pace and spin to create difficult, disruptive returns.

Materials and surface structures that reduce frictional coupling enhance the capacity to disrupt or neutralise incoming spin but limit the ability to actively generate new pace and spin. Conversely, high-friction surfaces support active stroke production and pressure imposition, at the expense of increased sensitivity to incoming spin.

Whilst not explicitly plotted in Fig. 2.3, dwell time (t_d) serves as a governing parameter across this spectrum: softer, thicker sponge increases dwell time; hardbat minimises it.

2.2.2 Functional Domains and Interrelationships

The mechanical behaviour of a bat may be understood through three interconnected domains (Fig. 2.4):

1. **Mechanical properties** are the intrinsic physical attributes determined by material composition and structural design, including blade geometry, lamina thickness, total mass, surface hardness, flexural stiffness, coefficient of restitution (CoR) and coefficient of friction (μ).

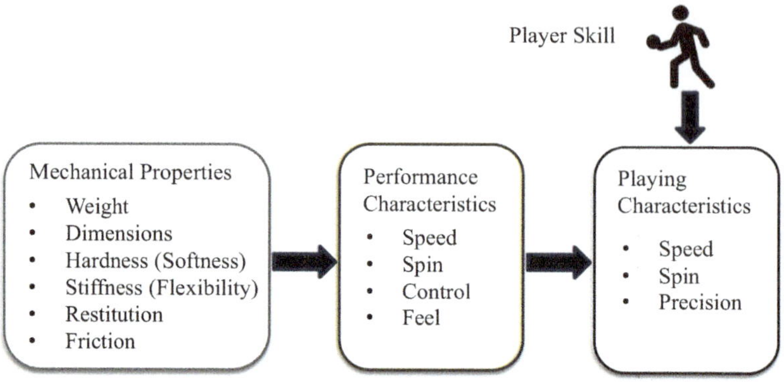

Fig. 2.4 Systems-hierarchy flowchart illustrating the causal progression from material properties to tactical outcomes. Mechanical properties such as mass and friction, determine performance characteristics such as speed and spin. Mediated by player skill, these characteristics shape playing characteristics and tactical effectiveness

2. **Performance characteristics** refer to the phenomenological, measurable playing manifestation of mechanical properties in resultant ball speed, spin, control and vibrational feedback (haptic and acoustic response, Sect. 2.4.3). Speed depends primarily on structural stiffness and sponge elasticity. Spin is governed by frictional traction and effective dwell time (t_d). Control reflects the linearity and reproducibility with which kinematic input is translated into projectile output.
3. **Playing characteristics** are the realised technical and tactical outcomes: the effectiveness of stroke execution (topspin loops, drives, blocks, pushes and chops), and the reliability with which tactical parameters (placement, rhythm and variation) are sustained under competitive conditions.

These three domains form a causal progression from mechanical properties to playing characteristics. Mechanical properties define the equipment's physical potential; performance characteristics express that potential in quantifiable ball behaviour; playing characteristics are shaped by the interaction between equipment capability and player proficiency. Establishing this structured ontology (Fig. 2.4) prevents the conflation of subjective feel with objective mechanics and provides a disciplined basis for equipment selection and technical development.

2.3 Hardness of a Table Tennis Bat

2.3.1 *The Problem of Proprietary Hardness Scales*

Hardness is one of the most consequential parameters in bat selection, yet it is also one of the most inconsistently reported. Manufacturers currently employ a range of proprietary hardness scales, which significantly complicates objective comparison of equipment components. European brands may report Shore A Durometer values or internal hardness indices, whilst Chinese manufacturers frequently use degree-based ratings (e.g. 39°) that correlate only loosely with Shore A measurements. Japanese manufacturers often provide relative performance matrices, commonly expressed on scales such as 1–10, characterising attributes including speed, spin and control. Whilst useful within a single product catalogue, these ratings are not directly comparable across manufacturers.

Selecting equipment based on such varied descriptions is therefore equivalent to judging a violin by its aesthetic finish rather than the density of its plates or the arching of its body, where slight geometric variations transform the instrument's resonance. This variability underscores the requirement for a systematic, engineering-led approach to material specification.

2.3.2 *Defining and Standardising Hardness Measurement*

Hardness (H) is defined as a material property describing resistance to localised plastic deformation such as indentation or scratching. It is conceptually and measurably distinct from stiffness, which describes the elastic deflection of the entire structure under load (Sect. 2.4).

Fundamentally, hardness is expressed as the ratio of applied normal force (F_n, in Newtons) to the resulting contact area (A, in m^2) of the indentation:

$$H = \frac{F_n}{A} \tag{2.1}$$

Although standardised modern hardness tests such as the Shore A method quantify indentation depth rather than direct area measurement, Eq. (2.1) captures the governing physical principle.

A harder rubber deforms less upon impact, creating a smaller contact patch and a reduced dwell time (t_d), resulting in a more direct, high-velocity rebound. A softer rubber undergoes greater elastic deformation, increasing contact area and prolonging t_d, facilitating greater surface traction and enhanced spin generation.

Two primary engineering measures are used to quantify component hardness:

Table 2.2 Typical hardness values of common blade woods and sponge–rubber combinations

Material	Elastic modulus (GPa)	Janka hardness (N)
Balsa (very soft, light)	3.0–3.7	~300
Ayous (Obeche)	~6.5	~1,960
Pine (soft)	8.5–10.0	1,690–1,870
Limba	~10.5	~2,990
Pine (hard)	11–12	3,290–4,120
Rubber + Sponge Combination	*0.01–0.10*	*Shore A 30–55*

- **Shore A durometer** measures indentation resistance in soft polymers (sponge and rubber) on a 0–100 scale. Typical values for table tennis components range from approximately 30 (very soft) to 55 (relatively hard).
- **Janka hardness test** measures the force required to embed an 11.28 mm steel sphere to half its diameter into wood. Values for common blade woods range from approximately 300 N for balsa to over 4,000 N for hard pines (Table 2.2).

The elastic modulus (i.e. Young's modulus) measures the stiffness of a material, and it is an *intrinsic* property.

2.3.3 Selecting Blade and Rubber Hardness

With hardness now defined and measurable, a rational basis for equipment selection becomes possible, replacing marketing language with quantifiable mechanical parameters. Manufacturers combine woods of differing Janka hardness with rubbers of specified Shore A values to generate a spectrum of hardness, stiffness and energy transfer characteristic.

A representative five-ply, all-wood construction illustrates how these parameters are distributed structurally:

- **Core**: high-compliance material (e.g. ayous), ~3.6 mm
- **Intermediate plies**: medium-hardness material (e.g. limba), ~0.4 mm each
- **Outer plies**: relatively harder material (e.g. limba), ~0.4 mm each
- **Total thickness (*h*)**: ~5.2 mm
- **Blade weight**: ~87 g

Such a construction provides moderated hardness and flexural stiffness with distributed compliance across the blade thickness. When paired with medium-soft rubbers (1.4–1.7 mm; Shore A 32–36), the system exhibits relatively linear rebound behaviour and extended dwell time, characteristics particularly suitable for foundational skill development and broadly accessible play.

More advanced players often refine equipment selection through systematic experimentation, informed by an understanding of how material hardness and structural stiffness influence ball response. The blade (weight, ply count, handle shape and Sect. 2.4) forms the stable core of a setup and may last for a decade; comfort in the hand is the first criterion. Rubbers are consumables, usually replaced anywhere from after a single competition to every two years.

This replacement cycle allows controlled testing. The principle is simple: change one variable at a time. Test a new rubber on the same blade to isolate its effect. When replacing both sides, change one side first and test it in matches before changing the other. Match play reveals responses that practice may not. Changing both rubbers at once makes it difficult to identify which change affected performance.

- **Greater outer-ply hardness**: Increasing ply count or incorporating high-modulus composites (e.g. carbon fibre) generally raises structural stiffness, shortens contact duration (t_d) and increases rebound velocity. Harder rubbers (Shore A 40+) deform less under moderate loads and return energy more directly. Under sufficiently large and concentrated impact, however, the rubber and sponge can be compressed more fully, increasing ball engagement and permitting high spin generation without sacrificing control.
- **Greater blade compliance**: Softer outer plies and lower structural stiffness generally extend dwell time. Softer rubbers deform more readily and can increase contact conformity and frictional traction. If the rubber–sponge system is too soft for the applied load, however, it may bottom out, causing an abrupt loss of control and reducing the capacity for consistent spin generation.

The material values in Table 2.2 and the thickness ranges in Fig. 2.1 are compiled from published material data, product specification and common blade constructions. They represent a typical, carefully selected configuration and provide a practical reference for both analysis and applied use.

2.4 Structural Stiffness and Vibro-Acoustic Behaviour

Stiffness (k) is defined as a structure's resistance to elastic deflection under an applied load. For a table tennis blade, structural stiffness is the primary determinant of rebound velocity, effective dwell time (t_d) and vibro-acoustic response. It is through this last dimension that the player perceives, in real time, the mechanical quality of each stroke.

In its simplest form, stiffness is

$$k = \frac{F}{\Delta} \tag{2.2}$$

where Δ (m) is the elastic deflection resulting from an applied force F (N).

2.4.1 Blade Stiffness, the Cube Law and Flexural Rigidity

Blade stiffness may be approximated using a cantilever beam model, where the handle acts as a fixed support and a normal load is applied at the blade head. Under this assumption (Hibbeler 2018), the static elastic deflection is

$$\Delta = \frac{FL^3}{3eI_A} \tag{2.3}$$

where:

- L = the blade head length (~15.2 cm)
- e = the elastic modulus (~10 GPa for wood; Table 2.2)
- I_A = the second moment of area (m^4), accounting for the cross-sectional geometry

For a rectangular cross-section,

$$I_A = \frac{Wh^3}{12} \tag{2.4}$$

where:

- W = the blade width ($\sim$ 15.0 cm),
- h = the blade thickness ($\sim$ 5.2 mm).

Substituting Eq. (2.4) into Eq. (2.3) gives

$$\Delta \propto \frac{4FL^3}{eWh^3} \tag{2.5}$$

Deflection is therefore inversely proportional to h^3. Since $k = F/\Delta$, stiffness scales with the cube of thickness:

$$k \propto h^3.$$

Beam Stiffness ($e{\cdot}I_A$) and Plate Rigidity (D)

For the cantilever beam, the governing stiffness term is ($e{\cdot}I_A$), combining material property (e) and cross-sectional geometry (I_A), and thus describing resistance to global bending. Carbon fibre possesses a substantially higher modulus than natural wood, reflecting its greater *intrinsic* material stiffness.

However, the blade head is more accurately modelled as a thin plate. Its intrinsic bending resistance per unit width is described by the *flexural rigidity* D (Leissa 1969):

$$D = \frac{e \cdot h^3}{12(1 - \upsilon^2)} \tag{2.6}$$

where υ is Poisson's ratio (typically 0.25–0.35 for wood).

The relationship between the two formulations is as follows:

- Beam stiffness: $k = e\frac{Wh^3}{12}$ (global cantilever bending, a rectangle cross-section)
- Plate flexural rigidity, $D = \frac{eh^3}{12(1-\upsilon^2)}$ (local bending per unit width)

Both scale with h^3 – the cube law. The differences are:

- Width W appears only in k
- Poisson's ratio υ appears only in D

Whether considering global deflection (k) or local bending (D, vibration), the dominant geometric scaling is identical: doubling thickness increases stiffness eightfold, and a 10% increase in thickness yields approximately a 33% increase in stiffness, effects that far exceed variations in elastic modulus. This cube law is the principal geometric mechanism in blade design, making thickness the most consequential design variable.

Length and width are secondary but not negligible. Deflection scales cubically with length (L^3Eg. (2.3)), so increases in effective length substantially raise flexibility and modify vibrational behaviour (Sect. 2.4.3). In practice, commercial head lengths are tightly standardised (15.0–15.2 cm), limiting variation in this parameter (Fig. 2.6c). Effective length may, however, be adjusted through handle geometry (e.g., length L_l = 9–12 cm), providing an alternative means of tuning blade stiffness. Width enters linearly but varies only marginally (14.7–15.0 cm). In both theory and practice, thickness remains the dominant design parameter.

2.4.2 *Effective Stiffness of Blades*

For a multi-ply or composite blade, the structural behaviour is defined by its effective flexural stiffness $(e \cdot I_A)_{eff}$, calculated using the Parallel Axis Theorem (Hibbeler 2018):

$$(e \cdot I_A)_{eff} = \sum (e_i \times I_{Ai} + e_i A_i d_i^2) \tag{2.7}$$

where

- e_i: the Young's modulus of ply i
- I_{Ai}: the second moment of area of ply i about its own centroid $(\frac{Wh_i^3}{12})$
- A_i: the cross-sectional area of the ply $(W \cdot h_i)$
- d_i: the distance from the ply's centroid to the neutral axis

The $d_i{}^2$ term is dominant: stiffness increases with the square of the distance from the neutral axis. Placing high-modulus materials such as carbon fibre near the outer surfaces therefore maximises their contribution to overall structural rigidity.

Consider a representative three-ply blade:

- **Core**: ayous, $h_0 = 3.6$ mm, $e \approx 4.0$ GPa
- **Out plies**: limba, $h_1 = h_{-1} = 0.8$ mm each, $e \approx 6.5$ GPa
- **Total thickness**: $h = 5.2$ mm.

A calculation using Eq. (2.7) shows that more than 75% of the total bending stiffness arises from the outer plies. The core contributes comparatively little direct bending resistance; its primary role is geometric, increasing d_i and thereby amplifying the stiffness contribution of the outer layers.

2.4.3 *Vibro-Acoustic Signature*

Whilst static stiffness determines deflection resistance, the player's perception is governed by the dynamic, multi-modal response during impact – the *vibro-acoustic signature*: a coupled frequency–amplitude spectrum transmitted through the handle (haptic) and radiated as sound (acoustic).

Structural Dynamics: The Blade-Rubber Assembly

For a thin isotropic plate (Leissa 1969), the *fundamental natural frequency* (f_1) scales approximately as:

$$f_1 = \frac{\lambda^2}{2\pi L^2}\sqrt{\frac{D}{\rho h}} \approx \frac{h}{L^2}\sqrt{\frac{e}{\rho(1-\nu^2)}} \tag{2.8}$$

where

- λ is a modal constant determined by the boundary conditions (how the handle holds the blade head).
- L is the blade head length.
- ρ is the material density.

The approximation absorbs the modal constant λ^2, geometric factors ($2\pi\sqrt{12}$), and isolates the governing dependencies as:

- $f_1 \propto h$: increasing thickness h raises the frequency f_1.
- $f_1 \propto 1/L^2$: increasing length L reduces the frequency f_1 quadratically.
- $\sqrt{e/(\rho(1-\upsilon^2)}$: captures the material contribution.

A modest increase in blade length, common in defensive designs, therefore, produces a marked reduction in f_1, explaining the perceptual association with slower response and increased dwell time.

Experimental modal analysis (Lionel et al. 2014) demonstrates that while theoretical models assume idealised orthotropic plates, the inclusion of rubbers adds mass and damping, shifting the frequency spectrum toward lower values without altering the fundamental mode shapes. Fast (OFF/OFF+) blades exhibit fundamental frequencies between 1,000 and 1,200 Hz, whereas defensive blades lie in 800–900 Hz. These flexural modes (~800–1,200 Hz) dominate the haptic channel, providing the primary tactile cues for stroke crispness and dwell time. The accompanying acoustic response serves as an audible proxy for mechanical impact during the contact: high-frequency transients associated with ball-rubber 'snap-through' correlate with higher coefficients of restitution, while lower-frequency flexural modes indicate greater energy dissipation and prolonged dwell time.

The Dual-Source Acoustic Model

The sound produced during play arises from a coupled interaction between the blade-rubber assembly and the ball. Russell (2018) identifies three distinct acoustic components:

- **Suppressed ball resonance (~5.9 kHz, Mode 2)**: Although the ball has a fundamental shell resonance at approximately 5.9 kHz, this mode is largely suppressed during contact. The relatively long impact duration (~1.8 ms) damps the ball's intrinsic vibration before it can radiate effectively, redirecting the system's acoustic energy into the blade-dominated frequency range.
- **Snap–through transients (8–12 kHz)**: At high-velocity impacts (> 5 m/s). The ball's thin-walled shell undergoes rapid non-linear buckling and recovery, a phenomenon known as 'snap-through' (Fig. 2.5b). This produces the high-frequency 'crisp' accent that players associate with offensive power.
- **Radiated assembly modes (1–2 kHz)**: These are the acoustically dominant components. Because the bat's surface area is substantially larger than the ball's, the blade-rubber assembly radiates sound far more efficiently into the surrounding air. These frequencies constitute the system's characteristic 'voice' (Fig. 2.5a).

A summary of the key frequencies is provided in Table 2.3.

a. 800–1200 Hz haptic channel (dwell time) and 1–2 kHz primary auditory channel (air radiation, bat voice)

b. > 8 kHz auditory channel, transient snap-through (air radiation, ball crisp accent)

Fig. 2.5 Dual-channel vibro-acoustic signature of table tennis impact. Impact generates both low-frequency vibration perceived through touch and higher-frequency acoustic response perceived as sound

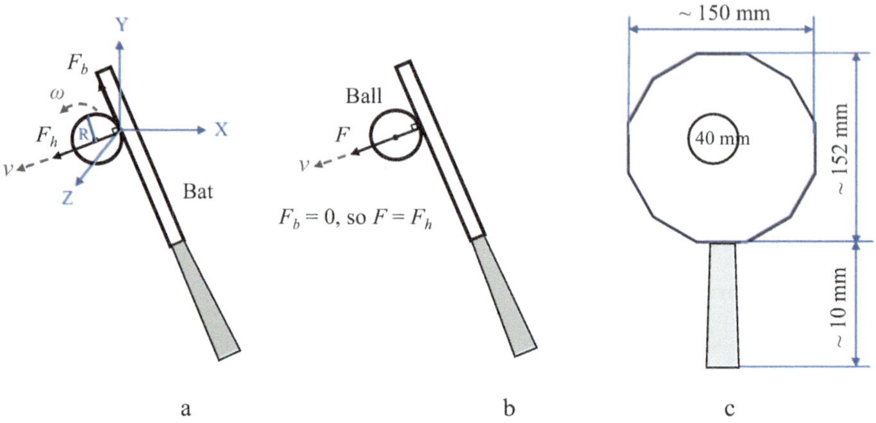

Fig. 2.6 Dual-Force Model and geometric scale. **a** Decomposition of the resultant contact force into normal (F_h) and tangential (F_b) components during ball–bat impact. **b** Limiting case of pure normal impact, where $F_b = 0$ and $F = F_h$. **c** Front-view dimensional comparison of a standard bat head and a 40 mm ball shown at accurate relative scale

Table 2.3 Vibro-acoustic spectrum components and their perceptual correspondences

Component	Frequency range	Dominant perception	Physical origin
Blade flexural modes	800–1200 Hz	Haptic (Dwell, vibration)	Cantilever plate bending
Blade–rubber assembly modes	~1–2 kHz	Audible (primary 'click')	Blade–rubber composite vibration
Ball natural resonance	~5.9 kHz	Suppressed during contact	Spherical shell vibration
Snap-through transient	8.5–12.5 kHz	Audible high-frequency 'crisp' accent	Non-linear buckling and recovery

Functional Interpretation

'Feel' is the combined effect of touch and sound. Touch indicates how much energy is transferred and how long contact lasts; sound indicates how hard the ball is struck. The 'click' contains two sounds, a low-frequency tone (1–2 kHz) and a high-frequency spike (> 8 kHz). Understanding this dual-source coupling enables the bat to be adjusted in a targeted way to achieve the desired feel and sound.

2.5 Mechanics of Striking a Table Tennis Ball

The structural and material properties established in Sect. 2.4 determine what the bat can do; the *Dual-Force Model* describes what the player does with it. The mechanics of a ball–bat collision may be formalised through hitting and brushing forces, allowing equipment properties and player input to be linked directly to ball behaviour. This dual-force formulation is developed here as a mechanically sufficient abstraction that extends naturally into tactics and stroke production. It is consistent with nonlinear impact-analysis approaches to bat restitution, including the reduced-mass methodology used by Kawazoe and Suzuki (2010), while serving a distinct purpose: connecting contact mechanics with stroke construction and tactical function.

2.5.1 The Dual-Force Model

Every table tennis stroke may be decomposed into two simultaneous vector force components acting during the short dwell time (t_d).

Normal (or hitting) component (F_h) acts perpendicular to the bat surface, passing through or near the ball's centre of mass (Fig. 2.6). This component generates the translational velocity v (*m/s*) and determines forward trajectory.

Tangential (or brushing) component (F_b) acts parallel to the bat surface, generating angular velocity ω (*rad/s*) or spin through the applied torque, $T_{q1} = R \times F_b$.

The resultant contact force (F) is the *vector* sum:

$$F = F_h + F_b \tag{2.9}$$

The resulting ball state is defined by simultaneous linear and angular velocities:

$$U = (v, \omega) \tag{2.10}$$

2.5.2 The Role of Material Deformation

In an ideal rigid-body scenario, a pure normal force passing exactly through the ball's geometric centre would produce no spin (Fig. 2.6b). In practice, impacts involve significant elastic deformation of both the ball and the layered rubber-sponge-blade structure (Fig. 2.7), which governs spin generation through two principal mechanisms:

1. **Enhanced tangential traction**: deformation increases the effective contact area, raising frictional traction and improving the transmission efficiency of the tangential force (F_b).

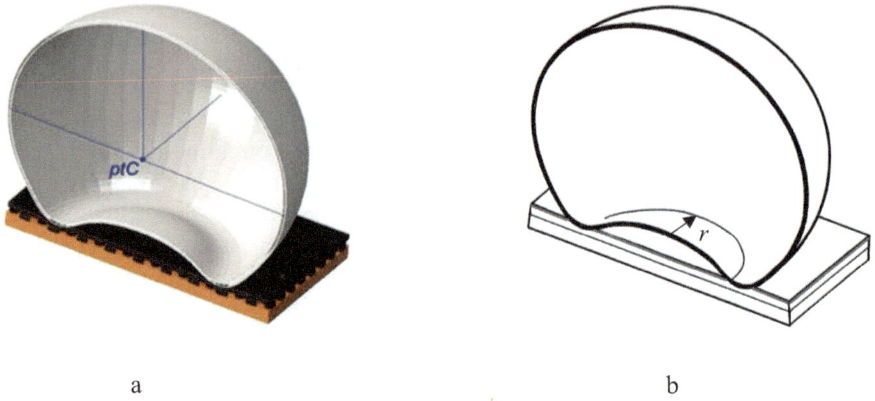

Fig. 2.7 Three-dimensional rendering of ball–bat impact. This figure illustrates elastic deformation of the ball and compression within the layered rubber–sponge–blade system. **a** Reproduced with permission from Rinaldi et al. (2016). **b** Simplified geometric representation showing the deformation radius r

2. **Eccentric normal-force contribution:** even when the hitting force (F_h) is nominally centred, the deformation shifts the effective contact centroid away from the ball's geometric centre, generating a secondary torque (Fig. 2.7b):

$$T_{q2} = r \times F_h \tag{2.11}$$

where r is the effective lever arm. This parasitic torque augments total angular velocity beyond that generated by F_b alone.

Nakashima et al. (2009) modelled how the rubber's elastic storage capacity drives significant deformation, with the stored potential energy acting as a dominant factor in the rebound, effectively 'returning' energy to the ball and influencing the exit angle and spin reversal. High-speed videography records ball velocities exceeding 30 m/s and spin rates of 100–150 rps (revolutions per second) during high-intensity attacking strokes. To the observer, the stroke appears singular and explosive. Mechanically, it is the controlled vector combination of the two forces. The challenge of skilful play is managing these forces throughout the dwell time.

Excessive normal force fully compresses the sponge, causing the ball to bottom out against the blade and producing an uncontrolled rebound with negligible spin. Excessive tangential force produces a 'thin' contact in which the ball skims the topsheet without engaging the sponge, yielding some spin but insufficient dwell time for consistent stroke production.

2.6 Impulse and Stroke Construction

Force describes an instant; impulse describes the duration of contact. The ball–bat interaction is a transient event lasting only a few milliseconds (dwell time t_d; Table 2.12). Within this narrow interval, the player modulates both the magnitude and orientation of applied forces to induce deformation and stability. The ball's post-impact state is governed not by instantaneous force magnitude alone, but by *impulse (J)*: the time-integrated effect of force over the contact duration.

Impulse determines change in both linear and angular momentum. It therefore distinguishes strokes that may appear similar in peak force but differ fundamentally in delivery, for example, a brief, concentrated strike; a sustained forward drive; or a spin-dominant brushing action. What matters mechanically is how the contact force is distributed over time.

Defining Impulse

Impulse is defined as:

$$J = \int_0^{t_d} F(t) \cdot dt \approx F_{avg} \times t_d \tag{2.12}$$

where:

- J = impulse (N·s)
- F_{avg} = average net force during contact
- t_d = dwell time (typically 0.6–1.8 ms, Table 2.12).

Newton's second law defines force as the time derivative of linear momentum (p):

$$F = \frac{dp}{dt} \tag{2.13}$$

or, more generally,

$$F = \frac{d(mv)}{dt} = m\frac{dv}{dt} + v\frac{dm}{dt} = ma + \dot{m}v \tag{2.14}$$

For a table tennis ball, mass m is constant ($\frac{dm}{dt} = 0$), so $F = ma$. The general form applies to variable-mass systems such as rockets, which expel propellant continuously.

Integrating over the contact interval yields the *impulse–momentum theorem*:

$$J = \int m\frac{dv}{dt}dt = m\Delta v = \Delta p \tag{2.15}$$

or equivalently,

$$J = m(v_f - v_i) \tag{2.16}$$

Impulse is therefore the change in momentum. Stroke outcome is determined by the accumulated impulse required to transform incoming velocity v_i into outgoing velocity v_f. To alter trajectory or motion, the applied impulse must exceed and redirect the ball's existing momentum.

2.6.1 Normal and Tangential Impulse Components

The dual-force model leads directly to two orthogonal impulse components.

1. **Normal impulse (J_h)**

Derived from the normal force F_h, this impulse drives the ball into the rubber, creating elastic deformation that increases effective contact area (Fig. 2.8). Deformation promotes frictional engagement and enables spin transfer.

2. **Tangential impulse (J_b)**

Derived from the tangential force F_b, this impulse generates spin through the *rotational* impulse-momentum relation:

$$J_b \times R = I_m \Delta\omega \tag{2.17}$$

where $I_m = \frac{2}{3}mR^2$ is the mass moment of inertia of the ball (a thin-walled hollow sphere), and $\Delta\omega$ is the change in angular velocity.

Increasing J_b directly increases angular velocity (ω), achieved through large brushing force (F_b) and prolonged dwell time (t_d).

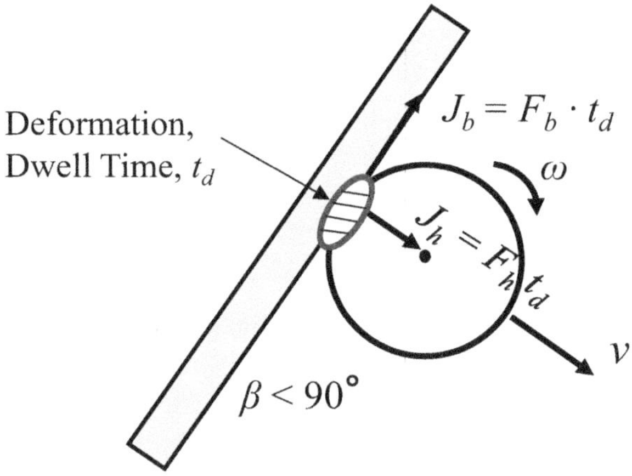

Fig. 2.8 Contact deformation patch and the impulse deformation. The highlighted contact patch denotes the region over which the total impulse is decomposed into normal impulse (J_h) and tangential impulse (J_b). Annotated formulas define $J = F \cdot t_d$, with resulting vectors for translational velocity (v) and spin (ω)

2.6.2 Impulse Profiles

Two strokes may produce identical spin rates but differ substantially in tactical effect due to their impulse profiles. *Stroke quality* therefore lies not in force magnitude but in how impulse is concentrated in time.

Traditional coaching heuristics such as 'wrapping around' or 'biting into the ball' describe sustained deformation and tangential traction during contact. In mechanical terms, these expressions refer to efficient impulse accumulation, the deliberate extension of the force–time integral rather than a brief strike. The heuristic and the physics describe the same event at different levels: one as playing feel, the other as contact mechanics.

This convergence points to a broader possibility. A player who internalises the concept of impulse does not discard the heuristic; they inhabit it more precisely. Equally, a coach who asks, "are you playing impulse or just force?" offers the player a question that is simultaneously physical and executable. The physics quantity becomes a performance cue. Dwell time is invisible; its consequences are not.

- **Explosive brush (speed-dominant loop)**: Characterised by high peak F_b and F_h over a moderate interval, producing a sharply concentrated *force–time* profile. Impulse is delivered rapidly, compressing the opponent's response window (t_w).
- **Sustaining brush (spin-dominant loop)**: Characterised by moderate forces applied over a longer interval, often facilitated by high-friction rubber. The impulse profile is smoother and more distributed.

These contrasting techniques are invisible to the naked eye and elusive to the beginner's touch. Their tactical divergence appears at the receiving end: the explosive stroke compresses response time through linear velocity, whereas the sustaining stroke forces adjustment to angular momentum and post-bounce spin.

2.7 Bat Angle and Grip Types

The *bat angle (β)* at impact determines which region of the ball is contacted and therefore constrains the type and magnitude of achievable speed and spin. Grip configuration, in turn, defines the natural range of attainable bat angles and shapes a player's default stroke tendencies. Bat angle thus forms the geometric bridge between mechanics and technique.

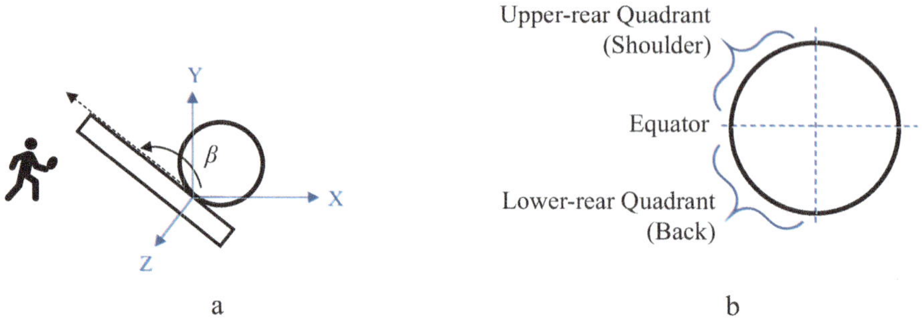

Fig. 2.9 Bat angle and ball contact regions. **a** Bat inclined at angle β relative to the ball. **b** Ball divided into geometric regions: *shoulder*, *equator* and *back*

2.7.1 Bat Angle and Contact Region

Let β denote the angle between the bat surface and the horizontal reference plane (X–Y; Fig. 2.9a). Although spin arises from tangential brushing action, bat angle determines which region of the ball is mechanically accessible (Fig. 2.9b).

Angle Classifications:

- **Neutral angle ($\beta = 90°$)**: the bat face is approximately vertical, exposing the *equatorial* region and facilitating sidespin when combined with a horizontal brushing vector (F_b).
- **Open angle ($\beta > 90°$)**: the bat face tilts backwards, exposing the lower-rear region of the ball and favouring backspin under downward–forward brushing action.
- **Closed angle ($\beta < 90°$)**: the bat face tilts forwards, contacting the upper-rear region (the shoulder) and favouring topspin under upward tangential impulse.

Effective technique involves continuous micro-adjustment of β to regulate the balance between normal and tangential impulse available at each contact point.

2.7.2 Penhold Essentials

The penhold grip predisposes the bat towards open bat angles ($\beta > 90°$) on both forehand and backhand sides, arising from the natural wrist flexion and vertical forearm alignment relative to the blade (Fig. 2.10).

Fig. 2.10 Penhold grip and open blade orientation. Forehand and backhand penhold grip configurations with schematic showing an open blade orientation ($\beta > 90°$)

Biomechanical Characteristics:

- **Default orientation:** the open resting angle provides direct access to the lower-rear region of the ball, facilitating backspin-oriented contact (e.g. short-range exchanges).
- **Compact wrist mechanics:** the grip permits rapid angular adjustment over small ranges, supporting short-contact strokes and variation in serve and receive.
- **Backhand topspin constraint:** achieving a closed angle ($\beta < 90°$) requires substantial wrist rotation, making traditional backhand topspin mechanically less natural.
- **Reverse Penhold Backhand (RPB):** pioneered at elite level by **Wang Hao**, **Xu Xin** and **Felix Lebrun**, the RPB employs the reverse face of the bat to achieve a closed-angle topspin configuration, expanding backhand topspin capacity and resolving the earlier geometric constraint.

These biomechanical tendencies account for the historical association of penhold with short-game control and variation, and the evolution towards more balanced topspin play that followed adoption of the RPB. Grip geometry, in this sense, shaped playing style before coaching did.

2.7.3 *Shakehand Essentials*

The shakehand grip introduces a functional asymmetry between forehand and backhand due to the neutral resting wrist position and differing forearm alignments (Fig. 2.11).

Biomechanical Characteristics:

- **Backhand:** the wrist extension required for backhand contact tends to produce an open angle ($\beta > 90°$), facilitating backspin pushes and controlled strokes. Achieving a closed angle for aggressive topspin requires coordinated wrist flexion and palm-down, inward rotation of the forearm.

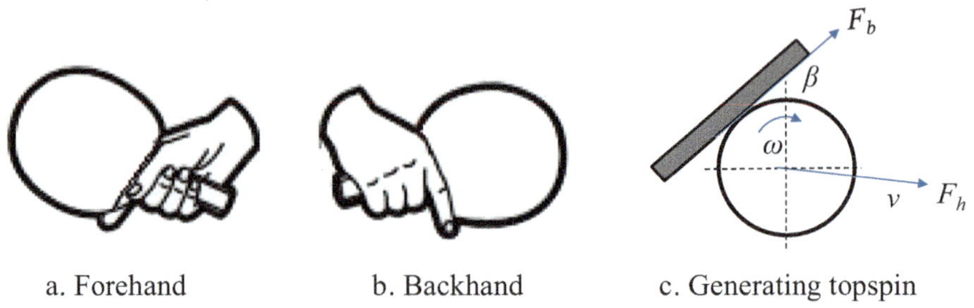

Fig. 2.11 Shakehand grip and bat-angle tendencies. Forehand and backhand grip configurations showing how the blade naturally closes ($\beta < 90°$) for topspin on the forehand and opens ($\beta > 90°$) for backspin on the backhand

- **Forehand:** neutral-to-flexed wrist orientation naturally produces a closed angle ($\beta < 90°$), favouring topspin generation. Opening the face for defensive backspin requires deliberate wrist extension (bending the hand backward).

This asymmetry explains common developmental patterns: forehand-dominant power combined with more control-oriented backhand play at amateur levels. At professional levels, deliberate training reduces but does not eliminate these inherent biomechanical tendencies. The forehand generally permits higher maximum spin and power, whilst the backhand offers faster attack initiation and more efficient transition into defensive control.

2.8 Friction in Ball-Bat Interaction

At the ball-rubber interface, friction is the dominant tribological parameter governing spin generation, contact-phase stability and outbound ball behaviour. Unlike intrinsic material constants such as hardness or stiffness, friction is an *emergent system property*: it arises from the interaction of surface chemistry, micro-texture, viscoelastic deformation and operating conditions, including contact force, sliding velocity and environmental factors such as humidity.

Understanding the interaction between these mechanisms is essential for linking equipment design to stroke mechanics and, ultimately, to tactical style. Varenberg and Varenberg (2012) further emphasised that frictional behaviour in table tennis should be characterised through objective physical parameters rather than subjective manufacturer ratings.

2.8.1 Coefficient of Friction

The static coefficient of friction μ_0 is defined as:

$$\mu_0 = \frac{F_f}{F_n} \tag{2.18}$$

where F_f is the maximum tangential, frictional force sustainable without macroscopic sliding, and F_n is the applied normal force.

This threshold determines whether the ball remains in stick contact or transitions into slip. Once sliding occurs, contact enters the kinetic regime characterised by $\mu_k < \mu_0$. Contact often alternates between stick and slip phases; the balance between these regimes governs spin transfer efficiency and rebound consistency.

Typical Friction Value Ranges

Under clean, optimal conditions, approximate static friction coefficients measured using quasi-static (slow) sliding tests are (Varenberg and Varenberg 2012):

- Tacky inverted rubbers: $\mu_O \approx 2.0$–2.2
- Short and long pimples rubbers: $\mu_O \approx 0.8$–1.5
- Anti-spin rubbers: $\mu_O \approx 0.195$ (Gossard 2026)

These values are indicative (obtained from various product information) rather than fixed, as friction emerges from surface and contact conditions.

1. **Factors influencing friction**

The effective friction coefficient μ_e depends on:

- **Material and surface properties:** rubber formulation, topsheet adhesion, micro-roughness, pimple geometry and surface cleanliness (dust, oils, contaminants),
- **Contact conditions** including local contact pressure (= normal force/deformation area), sliding velocity and contact temperature (affecting rubber compliance), and
- **Environmental and aging effects also play a role:** humidity affects adhesion, temperature alters viscoelastic response, whilst oxidation and the loss or migration of plasticisers modify surface properties over time (Wu et al. 2021).

2. **Velocity dependence of friction**

During impact, the effective coefficient μ_e reflects contributions from both adhesive and hysteretic mechanisms. Measurements by Inaba et al. (2017), obtained from ball–table interaction experiments, indicate an approximately linear dependence on contact-point velocity:

$$\mu_e = 0.0011 \times v_{cp} + 0.2526 \tag{2.19}$$

where v_{cp} is expressed in km/h.

Although derived from ball–table measurements, this relationship illustrates the broader velocity dependence of frictional behaviour in table tennis contacts.

Table 2.4 Representative coefficients of restitution for ball–table impact and friction coefficients (μ_e) for ball–bat impact

Shot type	*CoR*	μ_e
Serve	0.96	0.25–0.27
Smash	0.77	~0.34+
Drive	0.81	0.28–0.30

Increased sliding speed enhances viscoelastic hysteresis, elevating the effective friction during fast impacts. Representative measured values are provided in Table 2.4. These data illustrate the complementary roles of normal restitution (speed control) and tangential friction (spin control) across stroke classes.

3. **Microscopic origins of friction**

The macroscopic friction forces observed at the ball–rubber interface arise from two fundamental mechanisms (Inaba et al. 2017 and Renuad et al. 2019):

- **Adhesive friction**: shearing of transient molecular junctions formed at contact interface. This mechanism dominates static friction and explains the high μ_0 in *tacky rubbers.*
- **Hysteresis friction**: energy dissipation due to viscoelastic deformation as the ball traverses the rubber surface. This mechanism strengthens with sliding speed and dominates in kinetic regimes (Inaba et al. 2017).

The static coefficient (μ_0) is primarily adhesive in origin. Once sliding begins, adhesion diminishes and hysteresis becomes dominant, which is why tacky rubbers excel in low-speed spin generation but may exhibit reduced effectiveness at high sliding velocities.

2.8.2 Adhesion and Hysteresis in Friction

The tribological framework above permits classification of modern rubbers into two engineering types: adhesion-dominant and hysteresis-dominant systems (Table 2.5).

1. **Tacky/non-tensor rubbers (adhesion-dominant)**
    - **Surface mechanism**: high μ_0 arises from adhesive bond formation and rupture.
    - **Damping behaviour**: elevated internal damping reduces elastic rebound, producing a 'dull' (damped) tactile response and controlled energy return.
    - **Membrane stiffness**: a thick, hard topsheet behaves as a stiff membrane; sufficient normal impulse is required to engage the sponge layer.

Table 2.5 Comparative rubber mechanics

Rubber mechanics	Dominant physics	Tactile response	Preferred impulse profile	Playing feature
Tacky/ non-tensor	Adhesion; bond rupture	Damped (dull), robust	Sustained: longer t_d, arcing trajectory	Player-driven strokes; arm-swing and carrying movements (driving, slicing)
Tensor (mechanical)	Hysteresis; viscoelastic deformation	Elastic, reactive (bouncy)	Explosive: shorter t_d, rapid acceleration	Rubber-assisted power across explosive short-game actions (poking, punching) and high-velocity topspin loops

- **Velocity sensitivity**: at high sliding velocities, adhesive junctions cannot reform rapidly enough to track the motion, reducing the adhesive contribution and requiring compensation through a more *linear* stroke path or a more *closed* bat angle to sustain effective spin transfer.

2. **Tensor rubbers (hysteresis-dominant)**

   - **Mechanical grip**: friction arises primarily from viscoelastic deformation rather than surface tackiness.
   - **Elastic response**: low internal damping and high restitution produce rapid elastic recovery, enhancing rebound speed and dynamic energy return.
   - **Dynamic deformation**: microscopic surface compliance sustains friction under high sliding velocities, enabling consistent spin transfer even in high-speed exchanges.
   - **Temporal behaviour**: the rapid snap-back response supports short contact intervals and efficient impulse concentration, favouring high-acceleration strokes such as flicks, loops and quick counterattacks.

3. **Friction and impulse architecture**

The mechanical differences of the two rubber types produce distinct impulse architectures (Table 2.5). Adhesion-dominant rubbers sustain contact, carrying the ball through the interaction with longer dwell time and gradual impulse accumulation. They are well-suited to player-driven strokes, such as driving and slicing, where arm-swing mechanics and control are paramount.

Hysteresis-dominant rubbers, by contrast, concentrate impulse into a brief, high-intensity window driven by elastic recovery. Viscoelastic hysteresis helps maintain friction at high speed, extending effective stroke range from explosive short-game actions through to high-velocity topspin loops.

Equipment choice therefore encodes a preferred impulse strategy before the rally begins.

2.8.3 *Pips-Out Rubbers*

Pips-out rubbers modulate friction primarily through geometric architecture rather than surface adhesion (Varenberg and Varenberg 2012, Renaud et al. 2019).

- **Reduced real contact area**: the discrete pimple structure limits effective contact area, thereby reducing the maximum attainable static friction force.
- **Early stick-slip transition**: the lower frictional threshold promotes earlier transition from static to kinetic friction under tangential loading.
- **Spin attenuation:** because slip initiates earlier, incoming spin is only partially transmitted to the outgoing ball.

Short pips typically favour direct hitting and blocking due to reduced spin sensitivity. Long pips, are engineered to amplify variability in rebound angle and spin response, producing disruption through controlled instability.

2.8.4 *Anti-spin Rubbers*

Anti-spin rubbers are engineered to minimise friction through material formulation rather than geometric reduction. Gossard (2026) characterises these surfaces using a Coulomb friction model where the friction coefficient ($\mu \approx 0.20$) and coefficient of restitution are significantly lower than those of inverted rubbers:

- **Low-adhesive topsheet**: a smooth, non-tacky topsheet promotes sliding and suppresses static frictional engagement.
- **Compliant sponge layer**: a soft, high-damping sponge layer prolongs dwell time, enhancing control and rebound stability ($CoR \approx 0.53$, Gossard 2026) despite minimal tangential traction.

Unlike worn or contaminated rubbers, anti-spin surfaces provide a stable, predictably low-friction interface. Their behaviour is mechanically distinct from that of inverted rubbers. Rather than gripping the ball, the smooth surface promotes a sliding regime across a broad range of incident velocities. The resulting reduction in rebound momentum disrupts the rally, requiring rapid perceptual and tactical adjustment from the opponent.

2.8.5 *Gauging Rubber Performance and Reducing Waste*

Across all rubber types, performance degrades with use and players often replace rubbers based on subjective feel alone, leading to both performance variability and unnecessary waste. A practical measurement methodology addresses this directly,

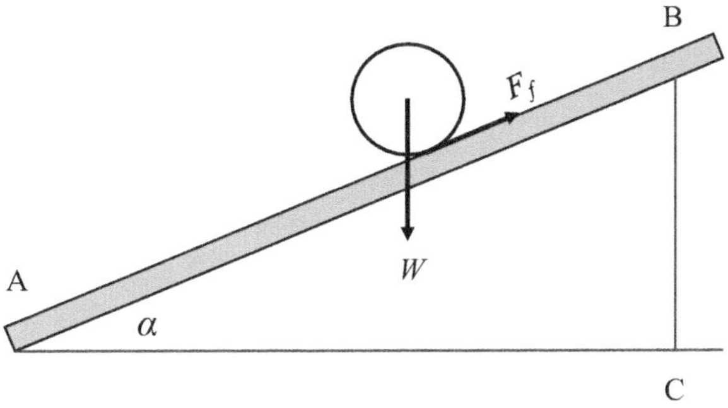

Fig. 2.12 An experimental setup for measuring the coefficient of friction. A ball rests on an inclined plane at angle α, with gravitational (W) and frictional (F_f) force identified

enabling objective quantification of rubber performance through direct measurement of the static friction coefficient (μ_0) against a standard ball (Fig. 2.12).

The measurement is based on the inclined plane principle. A linear actuator raises the plane from the point C until a ball placed on the rubber begins to roll. The static friction coefficient μ_0 is then calculated from the inclination angle (α) and the ball's acceleration (a):

$$\mu_0 = tan\alpha - \frac{a}{gcos\alpha} \tag{2.20}$$

where acceleration $a = 2s/t^2$ is derived from the travel distance (s) and time (t) between two points (A and B), measured, e.g., by light sensors.

This methodology demonstrates that rubber degradation can be objectively quantified with replicable measurements, transforming a subjective judgement into a measurable parameter. Such testing offers a practical pathway to optimise rubber replacement schedules, ensuring high-performance rubbers are reserved for competition whilst reducing unnecessary waste and promoting more sustainable equipment management across the sport.

2.9 The Game Event Sequence

The section introduces a compact structural notation (Table 2.6) that serves two purposes. First, it provides a causal description of rally dynamics. Second, it establishes a formal representation that supports both analytical modelling (Sect. 2.9.1) and computational implementation (Sect. 2.9.2). This dual role, descriptive and generative, positions the framework as a foundational layer for theoretical analysis and intelligent systems in table tennis.

Table 2.6 Game event sequence and notation

Notation	Event	Description	Governing physics
I (Impact)	Ball-bat contact	Dwell-time interaction during stroke execution	Force, impulse and friction
F (Flight)	Ball trajectory	Motion between contacts	Aerodynamic drag and Magnus forces
TC (Table Contact)	Bounce	Interaction with the playing surface	Restitution and friction

2.9.1 Structural Representation of a Rally

1. **Fundamental rally unit**

The smallest complete unit of player-to-player exchange is defined by:

$$I_1 \rightarrow F_1 \rightarrow TC \rightarrow F_2 \rightarrow I_2$$

This sequence captures the full causal chain, linking stroke execution, ball flight, table-surface interaction and opponent response (Table 2.6). The structure repeats recursively throughout a rally, forming the analytical backbone for temporal and spatial reasoning.

The rally sequence can be read as follows:

A player's bat makes first contact to the ball (Impact 1). The ball then leaves the bat and travels through its first flight (F_1), lands on the opponent's side of the table (Table Contact, TC), and rebounds into a second flight (F_2). The opponent then makes contact with the ball (Impact 2), completing the exchange. The next exchange begins:

$$I_3 \rightarrow F_1 \rightarrow TC \rightarrow F_2 \rightarrow I_4$$

Note that the recursive subscribes are applied to impacts, but not to flights or table contacts, as this is unnecessary. The number of the bounces (table contacts) can be inferred directly from the impact subscripts.

2. **Service case (initial bounce)**

The serve introduces an additional bounce event, as the ball must first strike the server's side:

$$I_1 \rightarrow TC_0 \rightarrow F_1 \rightarrow TC \rightarrow F_2 \rightarrow I_2$$

This preserves the same event logic whilst incorporating the regulatory requirement.

3. **Indexing and analytical scope**

Subscript indexing enables precise tracking of rally progression:

- **Impacts**: $I_1, I_2, I_3, \ldots$
- **Table contact if needed**: TC_0, $TC_{1,\ldots}$

This structure supports rigorous technical analysis and tactical pattern recognition: for example, identifying the third-ball attack (I_3) or fifth-ball strategy (I_5).

2.9.2 *Symbolic Representation for AI*

The sequence $I_1 \rightarrow F_1 \rightarrow TC \rightarrow F_2 \rightarrow I_2$ defines more than a descriptive model; it constitutes a formal, causal grammar of table tennis rallies. Because each transition is physically determined, the framework provides a structured substrate for artificial intelligence applications. It is not itself an AI system, but a representation upon which learning, inference and prediction can be built.

1. **Core representational properties**

The I–F–TC notation possesses several properties that distinguish it from flat event labelling (Table 2.7).

2. **Parametric embedding**

For computational use, each event in I–F–TC is associated with a parameter vector; several examples are provided in Table 2.8.

These vectors transform the game event grammar into a high-dimensional time series suitable for machine learning.

Table 2.7 Fundamental capabilities of the I–F–TC framework

Capability	Description
Causal grammar	Events are linked by physical necessity. Each impact defines the initial conditions of subsequent flight and bounce
Mechanical grounding	Each event maps directly to governing physics: impulse, restitution and friction (I), aerodynamics (F), restitution (TC)
Indexable structure	Subscripts (e.g. I_3) create persistent temporal anchors for recording and tactical analysis
Parameter extensibility	Events can carry arbitrarily rich parametric detail (Table 2.8) without altering the grammar
Compositional hierarchy	Sequences scale naturally: strokes → rallies → points → matches
Formal completeness	All legal rallies conform to the grammar; deviations correspond to faults or anomalies

Table 2.8 Parameter vectors for I–F–TC events

Capability	Parameter	Physical role
I (Impact)	Impact mechanics (F_h, F_b, J_h, J_b, CoR, μ), Trajectory angle α, initial velocity (v_0, ω_0)	Defines contact mechanics and launch conditions
F (Flight)	Trajectory spatial displacement (x, y, z), velocity (v, ω), spin ratio S, etc.	Describes spatial evolution and arc characteristics
TC (Table Contact)	Bounce location or placement (x, z); incident and exit velocity and angle	Determines post-bounce motion

3. **Implementation pathways**

The unified representation supports multiple classes of AI systems as summarised in Table 2.9.

4. **Structured sequence representation**

To enable computation, rallies must be encoded in a machine-readable format (e.g., indexed by rally identifiers and ordered event sequences),

$$\text{match\_id} \rightarrow \text{rally\_id} \rightarrow \text{exchange\_id} \rightarrow \text{I--F--TC}$$

Such sequencing is fundamental in data-driven analysis. And this event-level grammar provides the temporal scaffold for the later TTNS discretisation developed in Chapter 6, where impacts, trajectories and tactical states are further encoded into tokenised stroke and state representations suitable for AI system integration (Sect. 6.10).

Table 2.9 AI implementation pathways supported by the I–F–TC events

Implementation	Methodology	Output
Event detection	Computer vision + sequence models	Structured I–F–TC sequences from video
Pattern recogniser	Sequence modelling and mining	Tactical patterns (e.g. third-ball attack)
Outcome prediction	Transformer-based models	Win probability from partial sequence
Trajectory estimation	Physics-informed learning	Full state reconstruction
Generative simulation	Generative models	Synthetic rally sequences
Player profiling	Clustering and data mining	Strengths and behavioural tendencies
Tactical advisory	Reinforcement learning	Real-time decision support
Coaching feedback	Explainable AI	Interpretable technical feedback

2.10 Symbolic Notation for Spin

Spin is a defining tactical element of modern table tennis, underpinning both technical execution and decision-making. Without a formal representation, its manipulation cannot be analysed with the precision the game demands.

A standard Cartesian coordinate system (Fig. 2.13) is adopted to describe spin in a precise and operational manner.

- The Z-axis is parallel to the net.
- The X-axis is aligned with the direction of ball travel.
- The Y-axis is normal to the table surface.

Within the coordinate system, topspin, backspin and sidespin are represented using a compact symbolic vocabulary (Table 2.10) designed for rapid spin-axis recognition and tactical analysis. Symbol choices are both mnemonic and visually suggestive: the reversed epsilon (∂) for backspin, for instance, suggests backward rotation; the sigma (σ) for topspin suggests forward rotation; and the theta (θ) for sidespin captures the action directly, its horizontal line representing the equatorial brush across the ball's equator. Compound notation permits the concise representation of complex rotational states (Fig. 2.14): $\sigma\theta$ for topspin-sidespin hybrids, or A$\theta\sigma$ for anticlockwise sidespin combined with topspin.

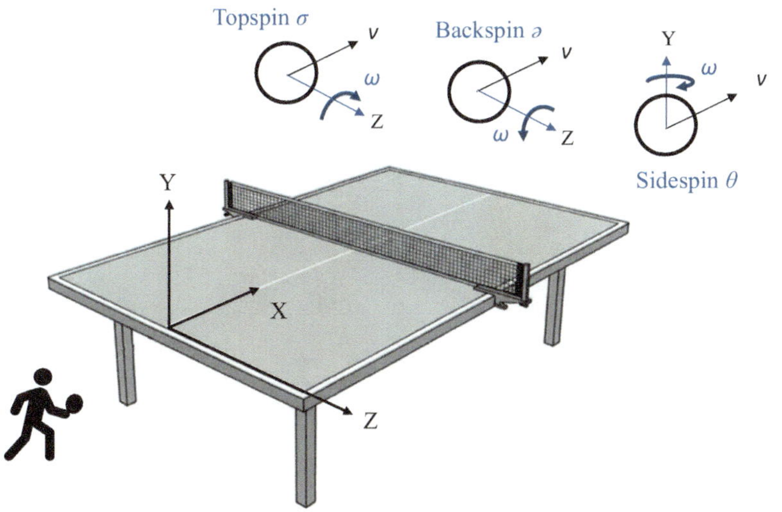

Fig. 2.13 Three-dimensional coordinate system and primary spin vectors in table tennis. The X, Y and Z axes define spatial reference frame for topspin (σ), backspin (∂) and sidespin (θ)

Table 2.10 Spin types and associated symbols

Spin	Symbol	Description
Topspin	σ	Forward rotation about the horizontal (Z) axis
Backspin	ɘ	Reverse rotation about the horizontal (Z) axis
Sidespin	θ	Rotation about a vertical axis (Y)
No-spin	χ	A deliberate absence of significant rotation
Compound spin	Combined	A synthesis of spin axes (e.g., A$\theta\sigma$ or L$\theta\sigma$ for anticlockwise or left sidespin + topspin)

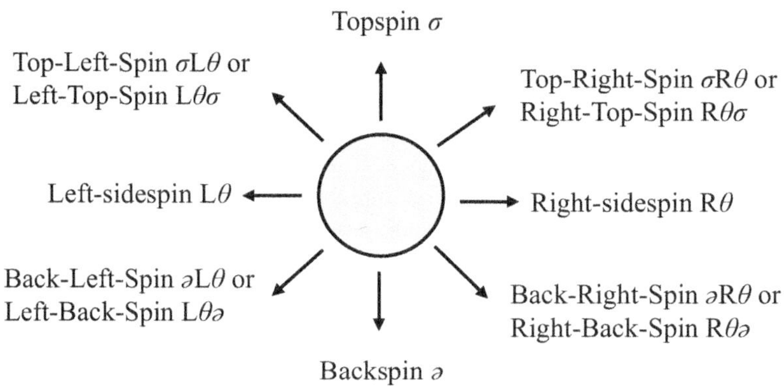

Fig. 2.14 Symbolic map of spin directions. A central circle representing a ball has eight arrows radiating outwards to represent different spin directions. Each direction is labelled with a name and a corresponding symbolic notation

2.11 Spin Ratio and Spin Transformation

The tactical identity of the ball is captured by the *spin-to-speed ratio*, or simply the *spin ratio*:

$$S = \frac{\omega \cdot R}{v} \tag{2.21}$$

where

- $\omega =$ angular velocity
- $v =$ translational velocity
- $R =$ ball radius ($R = 20$ mm for a 40 mm ball).

The forward velocity component is often the dominant contributor, particularly for topspin and backspin interactions. Thus, spin ratio can be more clearly defined as:

$$S_x = \frac{\omega \cdot R}{v_x} \tag{2.22}$$

where v_x is the speed component along the X axis (forward to the player, Fig 2.13). S_x represents the vertical spin ratio, i.e., topspin or backspin ratio, $S_{\sigma/ə}$. Previous studies have similarly identified the spin-to-speed relationship as a primary determinant of trajectory evolution, rebound sensitivity and spin-transfer behaviour during impact (Inaba et al. (2017), Rinaldi et al. 2019 and Gossard 2026).

S values typically fall well below unity because R is small, and translational velocities are large relative to angular velocities. Empirically, three regimes may be identified:

- $\mathbf{S \geq 0.7}$: spin-dominant, strong rotational influence; pronounced Magnus curvature and high post-bounce sensitivity.
- $\mathbf{S = 0.3–0.7}$: moderate spin, balanced translational and rotational contributions; tactically versatile.
- $\mathbf{S < 0.3}$: speed-dominant, light rotational influence; trajectory is flat, and pressure arises primarily from compressed reaction window (t_w). Light spin has little influence on tactics in terms of trajectory curvature or the acceleration and deceleration of rally pace.

These thresholds are not sharp boundaries but practical ranges, informed by measured ball velocities and spin rates observed in competitive play.

2.11.1 Spin Generation at Impact

Spin is generated when the bat applies a tangential force F_b to the ball during impact (t_d), producing a rotational impulse J_b that determines the ball's resulting angular velocity ω. The spin axis depends on the contact region of the ball (Fig. 2.9b):

- **Topspin (σ)**: Brush the *shoulder* of the ball.
- **Backspin ($ə$)**: Brush the *back* of the ball.
- **Sidespin (θ)**: Brush primarily across the *equatorial region*.

Compound spin is generated by combined vertical ($\sigma/ə$) and lateral (θ) brushing components (Fig. 2.14).

2.11.2 Aerodynamic Evolution during Flight

Spin does not remain static after impact. It evolves during flight (F_1), is transformed at table contact (TC), and is finally interpreted by the opponent during the post-bounce flight (F_2).

The two decay rates are fundamentally different in character. Translational velocity decelerates quadratically as drag force scales with v^2 (Eq. 3.2); fast balls shed speed rapidly, while slow balls coast. Rotational decay is gentler: the Magnus torque that brakes spin scales with $v \cdot \omega$ (Eq. 3.3), meaning that as v falls, the braking torque on ω falls with it. Spin decay therefore weakens.

This asymmetry governs the spin ratio $S = \omega \cdot R/v$. Differentiating with respect to time:

$$\frac{dS}{dt} = R\frac{v\frac{d\omega}{dt} - \omega\frac{dv}{dt}}{v^2}$$

Thus, the equations of motion during flight are

$$\frac{dv}{dt} = -K_d v^2, \ \frac{d\omega}{dt} = -K_m v\omega$$

where $K_d = \frac{1}{2}\rho A C_D$ (Eq. 3.2) and $K_m = \frac{1}{2}\rho A C_m$ (Eq. 3.3). Combining these equations yields the change in spin ratio during flight:

$$\frac{dS}{dt} = \omega R(K_d - K_m) \tag{2.23}$$

Thus:

- If $K_d > K_m$, $\frac{dS}{dt} > 0$, the spin ratio increases during flight.
- If $K_d = K_m$, $\frac{dS}{dt} = 0$, the spin ratio remains constant.
- If $K_d < K_m$, $\frac{dS}{dt} < 0$, the spin ratio decreases during flight.

For typical table tennis flight regimes, $K_d > K_m$ (Tables 3.1 and 3.2), the spin ratio increases during flight. Forward speed v_x decays under drag while spin remains approximately conserved, leading to a systematic increase in the forward spin ratio S_x. The ball arrives at the table slightly more spin-dominant than it left the bat (Inaba et al. 2017).

2.11.3 Spin Transformation at Table Contact

Figure 2.15 illustrates the frictional force F_f at table contact for topspin, backspin and sidespin.

1. **Backspin (∂): Braking and Dead Bounce**

For a backspinning ball, the lowermost surface moves forwards relative to the table at impact. The table therefore exerts a backward frictional force (Fig. 2.15b), reducing forward velocity whilst the associated torque opposes the rotation.

Depending on the magnitude of friction and the angle of incidence, three regimes may occur:

- **Partial decay**: backspin reduced but retained.
- **Full arrest**: rotation momentarily zero; forward motion heavily attenuated, producing the characteristic *dead* bounce.
- **True reversal**: the frictional impulse exceeds rotational inertia, and the ball exits with slight topspin, a behaviour often termed *ghost* bounce. Mohammed et al. (2019) demonstrate that this reversal results from extreme backspin coupled with frictional grip, which reverses the horizontal velocity vector.

In all three regimes, friction reduces tangential slip and redistribute energy between translation and rotation. Under typical contact conditions, the reduction in forward velocity Δv exceeds the reduction in angular velocity $\Delta\omega$, so the forward spin ratio S is reduced (Eq. 2.24). The post-bounce is slower and less spin-dominant, producing the braking behaviour that remains tactically demanding at I_2.

2. **Topspin (σ): Forward Kick**

For a topspin ball, the lowermost surface moves backwards relative to the direction of travel. The table applies a forward frictional force (Fig. 2.15a), converting part of the ball's rotational energy into forward translational motion – the characteristic *forward kick*. Friction simultaneously reduces spin, and the net effect under typical conditions is a reduction in spin ratio.

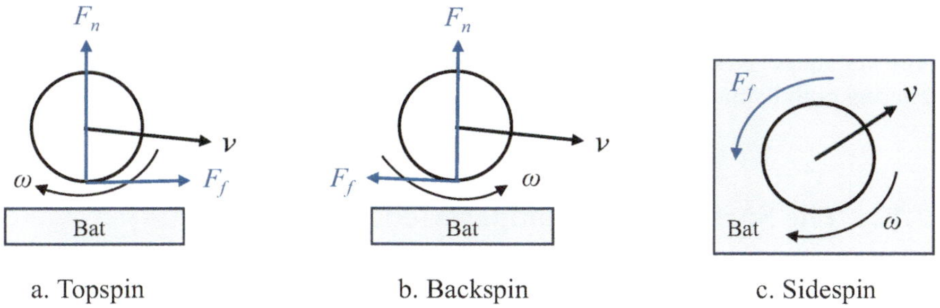

Fig. 2.15 Frictional scenarios at table contact. The diagrams show how topspin, backspin and sidespin modify friction, exit velocity and post-bounce trajectory

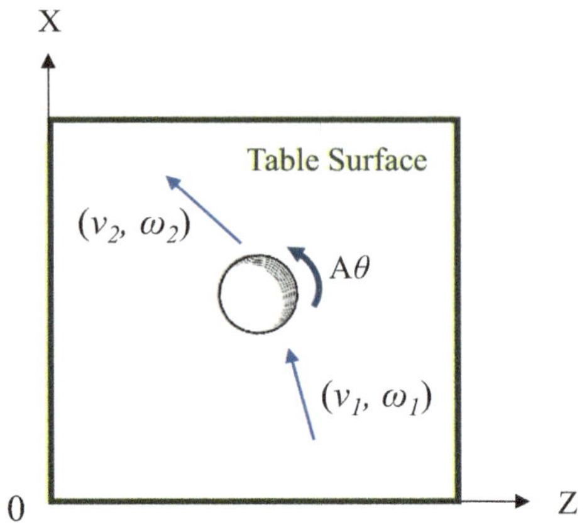

Fig. 2.16 Lateral path deviation under anticlockwise sidespin. Top-down X–Z plane view illustrating how anticlockwise sidespin ($A\theta$) alters the ball's velocity and trajectory after table contact

$$S_{post} < S_{pre} \tag{2.24}$$

The rebound becomes speed-dominant, compressing the opponent's reaction window (t_w) at I_2.

3. **Sidespin (θ): Lateral Deviation**

For a sidespinning ball, lateral slip at contact *introduces* a lateral frictional force (Fig. 2.15c). Forward velocity is reduced; a lateral velocity component is introduced (Fig. 2.16); and angular velocity is moderately attenuated.

Unlike topspin and backspin where redistribution occurs primarily along the forward direction, sidespin redistribution is primarily lateral: spin is reduced whilst lateral velocity is introduced. The resulting trajectory deviation produces a laterally misaligned approach at I_2, often perceived as unpredictable.

Summary: Bounce as Spin-Speed Redistribution

Across all spin types, table contact redistributes translational and rotational motion over a millisecond-scale bounce interval, driving the ball towards a rolling condition (Table 2.11). For topspin and backspin, the dominant coupling is between spin and forward velocity; for sidespin, redistribution occurs primarily in the lateral direction. The kinematic asymmetry introduced at contact reshapes the ball's state prior to I_2, altering both its trajectory and the receiving player's available response options.

2.12 The Spring-Mass-Damper Model

Kawazoe and Suzuki (2010) formulated a two-degree-of-freedom (2DOF) spring–mass–damper model to describe the impact force and deformation dynamics in table tennis. The thin-walled hollow sphere of the ball and the laminated sandwich structure

Table 2.11 Table contact as a spin transformer

Ball state	Change in v_x	Change in ω	Change in S_x	Tactical result
Backspin	Decrease	Moderate to large decrease	Decrease	Dead bounce
Topspin	Slight increase or moderate decrease	Moderate decrease	Decrease	Forward kick
Sidespin	Decrease (forwards) + lateral gain	Moderate decrease	Redistribution	Lateral deviation

Table 2.12 Technical evolution from hardbat to modern sponge construction

Feature	Hardbat	Sponge bat
System stiffness (k_1)	High (direct wood contact)	Variable, elastic composite
Damping (c_1)	Low (minimal energy absorption)	Moderate to high
Dwell time (t_d)	≈0.6–0.83 ms (Russell 2018)	≈1.3–1.8 ms (Russell 2018)
Energy mechanism	Predominantly dissipative	Elastic storage and restitution

of the bat lend themselves even more naturally to representation as two coupled spring–mass–damper subsystems during collision (Fig. 2.17).

Over the millisecond-scale contact interval, both bodies deform elastically, temporarily storing strain energy whilst dissipating part of it through internal damping. This dynamical interaction is expressed macroscopically through two directly measurable quantities: the dwell time (t_d) and the coefficient of restitution (*CoR*; Sect. 2.13).

Each subsystem is parameterised by:

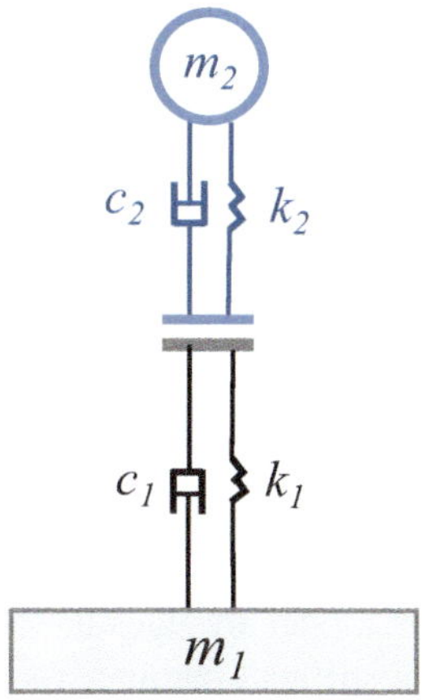

Fig. 2.17 Spring–mass–damper representation of ball–bat collision, showing mass (m), stiffness (k) and viscous damping (c) components for both subsystems

- **Ball (System 2)**: mass m_2, stiffness k_2, viscous damping coefficient c_2
- **Bat (System 1)**: effective mass m_1, stiffness k_1, viscous damping coefficient c_1

For lightly damped systems, damping is conveniently expressed through the dimensionless damping ratio ζ, related to the viscous coefficient by $\zeta = c/(2\sqrt{km})$.

2.12.1 Ball Stiffness and Damping

During impact, the ball may be modelled as being subjected to a concentrated point load. For small linear deformations, its stiffness is given by (Leissa 1969):

$$k_2 = \frac{4eh^2}{\sqrt{3(1-\upsilon^2)}R} \tag{2.25}$$

where:

- e is the elastic modulus of the ABS plastic material (~2–4 GPa)
- h is the wall thickness ($\approx$0.4 mm)
- R is the outer radius ($R = 20$ mm for a 40 mm ball)
- υ is Poisson's ratio ($\upsilon \approx 0.33$).

This gives an estimated range of $k_2 \approx 40 - 78$ kN/m.

The 2000 rule change, which increased the ball diameter from 38 to 40 mm, reduced shell stiffness k_2 by approximately 5%. Since stiffness scales inversely with radius, the larger ball is slightly more compliant, producing softer contact and a modest reduction in rebound velocity. These effects follow directly from thin-shell theory, Eq. (2.25), without requiring empirical observation.

Ball damping is commonly inferred from rebound tests against a rigid steel plate. With $CoR \approx 0.88$–0.93 (ITTF 2026), the corresponding damping ratio is $\zeta_2 \approx 0.025 - 0.04$, and the associated viscous coefficient is $c_2 = 2\zeta_2 \sqrt{(k_2\ m_2)} \approx 0.5 - 1.2$ N s/m. Variations in polymer composition, wall-thickness uniformity and manufacturing tolerances produce measurable differences in k_2 and damping behaviour that skilled players readily detect.

2.12.2 Bat Stiffness and Damping

The structural origins of bat stiffness were examined in Sect. 2.4.1. However, it is for the blade alone. The present model arranges the blade, sponge and rubber sheet in series, yielding an equivalent stiffness parameter k_1 for the entire bat:

$$\frac{1}{k_1} = \frac{1}{k_{blade}} + \frac{1}{k_{sponge}} + \frac{1}{k_{rubber}} \tag{2.26}$$

where:

- k_{blade}: derived from $(e{\cdot}I_A)_{\text{eff}}$ or D (Sect. 2.4.1)
- k_{sponge}: compliance of sponge, often dominant at low loads
- k_{rubber}: compliance of rubber.

The sponge is typically the most compliant element and dominates the response at low to moderate loads, giving $k_I \approx 30 - 80$ kN/m. Total bat mass is typically: $m_I \approx 0.17 - 0.19$ kg.

Energy dissipation arises primarily from viscoelastic losses in the sponge and rubber. This is represented by an effective damping ratio $\zeta_I \approx 0.10 - 0.20$, corresponding to $c_I \approx 0.6 - 1.2$ N s/m. These are effective impact parameters, not intrinsic material constants, and dependent on deformation amplitude, impact location and blade geometry.

2.12.3 Coupled System Dynamics

During impact, the ball and bat form a coupled dynamical system characterised by an equivalent stiffness k_e and reduced mass m_e:

$$\frac{1}{k_e} = \frac{1}{k_1} + \frac{1}{k_2}, k_e = \frac{k_1 k}{k_1 + k_2} \quad m_e = \frac{m_1 m_2}{m_1 + m_2}$$

Since $m_I \gg m_2$, the reduced mass is effectively governed by the ball: $m_e \approx m_2 \approx 2.65 \times 10^{-3}$kg.

For representative values $k_I = 5.5 \times 10^4$ and $k_2 = 6.0 \times 10^4$, the equivalent stiffness is

$$k_e \approx 2.87 \times 10^4 \text{ N/m}$$

giving a natural frequency:

$$\omega_n = \sqrt{\frac{k_e}{m_e}} = 3.29 \times 10^3 \text{rad/s} \approx 524 \text{ Hz}$$

The dwell time t_d, is approximated by the half-period of the coupled system:

$$t_d = \frac{\pi}{\omega_n} = \pi\sqrt{\frac{m_e}{k_e}} \approx 0.95 \text{ ms}$$

This is consistent with experimental observations of ~1 ms contact duration. Since $m_e \approx m_2$ is effectively fixed by the ball mass, t_d is governed predominantly by the effective stiffness k_e, and hence by blade construction, i.e., blade material, geometry, thickness and compliance of the rubber–sponge assembly.

2.12.4 The Catapult Effect

With the system parameters k_1 and k_2 established, the *catapult* (or *trampoline*) *effect* follows directly from the sponge–mass–damper framework. Elastic energy stored during compression (governed by k_1) is partially recovered during unloading, adding to the player's applied force.

Higher compression (as in loop or smash strokes) increases stored energy and rebound velocity. In low-force interactions, elastic recovery can dominate, producing a rebound stronger than the player's intended. The catapult effect is therefore neither intrinsically beneficial nor detrimental; it is a structural consequence of elastic compliance, whose influence depends on how it is used tactically.

2.12.5 Hardbat, Sponge and the Evolution of Impact Mechanics

The spring–mass–damper parameters established above provide a precise framework for understanding the most consequential equipment transitions in the sport's history. The shift from the *hardbat* era to the *sponge revolution* of the 1950s represents a fundamental change in system stiffness, damping and dwell time – summarised in Table 2.12. But the most sophisticated insight that this framework offers is not the contrast between hardbat and sponge but the possibility of their deliberate combination on a single bat.

The Hardbat Regime

Hardbat produces near-instantaneous impacts. Deformation is minimal, dwell time is short ($t_d < 1$ ms), and spin arises mainly from topsheet friction, a consequence of the bat's high stiffness k_1 and low damping c_1, which together limit elastic energy storage and return. Skilled players such as Marty Reisman (Sect. 2.2.1) compensated by manipulating impulse vector (J_h, J_b) through precisely directed driving, guiding or chopping motions that prolonged the effective contact interval and shaped the outgoing ball trajectory through timing rather than equipment compliance.

The Sponge Regime

Sponge layers introduce compliance, reducing effective stiffness and increasing dwell time. Under typical impact conditions with modern sandwich rubber, t_d lies in the range of approximaely 1–2 ms, with softer, thicker configurations approaching the upper bound. This elastic deformation enables energy storage and return, producing the catapult effect: part of the rebound energy originates from the bat itself rather than solely from the player's stoke.

The mechanical shift underlies the divergence of impulse architectures described in Sect. 2.8.2: adhesion-dominant rubbers favour sustained impulse generation through prolonged contact, whereas tensor rubbers exploit elastic restitution to deliver a concentrated impulse within a shorter contact window.

This compliance is highly sensitive to layer geometry; Tiefenbacher and Durey (1994) quantified that increasing sponge thickness from 1.3 mm to 2.1 mm yields a 10% increase in rotational velocity – an effect that, while seemingly marginal in geometric terms, straddles the threshold between controlled placement and the high-speed, mechanically 'uncontrollable' rebound characteristic of modern offensive equipment.

Synthesis: Dual Mechanical Regimes

The most sophisticated realisation of these principles is their combination on a single bat. **Deng Yaping**, widely regarded as one of the greatest female players in the sport's history and World No. 1 for eight consecutive years from1991, represents the definitive case study. Her equipment combined inverted rubber on the forehand with long pimples on the backhand.

The forehand operated as a high-friction, compliant system, generating heavy topspin and explosive smash production: the classic sponge regime delivering maximum temporal pressure. The backhand, by contrast, acted as a high-damping system. Long pimples deform and buckle under load, producing a low-speed return with disrupted trajectory and decoupled spin.

What distinguished Deng from conventional long-pimple users was her tactical exploitation of this mechanical decoupling. Most long-pimple players retreat from the table and use the spin-reversal passively, absorbing and redirecting pace. Deng remained close to the table and used the pimples actively, poking and punching with the backhand to produce flat, wobbling returns of unpredictable trajectory that forced the opponent to adjust and hesitate, setting up the forehand kill.

Era Periodisation

The evolution of these mechanical regimes maps onto three broadly recognised competitive eras:

Hardbat and Placement Era (to 1952): extended rallies governed by spin sensitivity and positional control. Exchanges formed a tactical dialogue in which the opponent's spin is utilised rather than overpowered. Placement and positional control were paramount.

Speed and Spin Era (1952 – c.2000): sponge rubber introduced heavy backspin and spin-reversal, dominant in the 1950s and 1960s. From the 1960s the loop drive responded with topspin, making the era a duel of backspin versus topspin, with speed and spin steadily increasing.

Top spin Era (c.2000 – present): extreme topspin, tight placement and temporal control, compressing the opponent's reaction window (t_w) through speed, trajectory curvature and variation in interception timing (τ).

This periodisation makes several iconic techniques visible and situates them within a broader landscape of technique, tactics and skill. It offers a practical map of table tennis development, allowing players at every level, from leisure players and competitive amateurs to Olympic athletes, to recognise where their own skill set lies and identify viable ways to win.

2.13 Restitution and Reactive Mechanics

2.13.1 Fundamental Restitution (CoR)

The coefficient of restitution (CoR) determines rebound speed, the perceived 'liveliness' of the bounce, and the passive energy-return behaviour of equipment during impact.

The CoR applies only to the normal component of the interaction. When the ball is hit harder, the percentage of energy returned decreases due to higher damping in the polymeric materials of the bat (Kawazoe and Suzuki 2010). Tangential behaviour, separately, governs the frictional mechanisms central to spin generation and transfer (Sects. 2.8 and 2.11).

For a ball striking a stationary surface, the coefficient of restitution is defined as the ratio of rebound velocity v_2 to incident velocity v_1, measured along the normal direction:

$$CoR = \frac{v_2}{v_1} \tag{2.27}$$

In controlled vertical drop experiments, the same quantity may be determined indirectly from rebound heights. Under negligible aerodynamic influence, the velocity ratio reduces to the square root of the ratio of successive peak heights:

$$CoR = \sqrt{\frac{H_2}{H_1}} \tag{2.28}$$

A perfectly elastic collision corresponds to $CoR = 1$, indicating complete recovery of normal incident velocity. A perfectly inelastic collision corresponds to $CoR = 0$, indicating total loss of recoverable normal motion. In practice, $0 < CoR < 1$ due to viscoelastic deformation, acoustic radiation and thermal losses.

The progressive decay of bounce height, a direct manifestation of $CoR < 1$, is vividly illustrated through numerical simulation. Figure 2.18 shows the decaying trajectory of a table tennis ball dropped from 707.2 mm, modelled in MATLAB

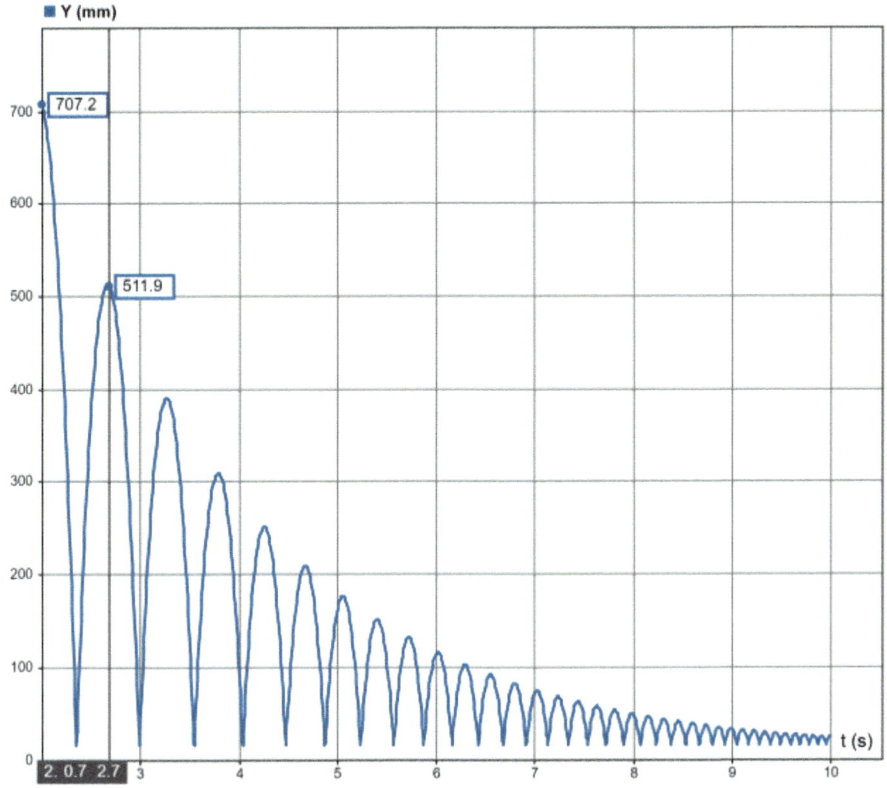

Fig. 2.18 Vertical position versus time, showing the diminishing bounce height of a ball dropped from 707.2 mm and illustrating height reduction over successive impacts. The plotted data are from the author's 2023 Simscape study of a freely falling ball

with aerodynamic drag included. Each successive peak is lower than the preceding one, reflecting energy dissipation at each impact. The simulation links microscopic viscoelastic damping at contact to observable macroscopic motion and permits practical evaluation of *CoR* from any pair of consecutive peaks using Eq. (2.28). As the amplitude decreases, flight time shortens, and the apparent bounce frequency increases slightly due to the reduced vertical excursion.

2.13.2 Table Restitution

International Table Tennis Federation (ITTF) regulations (ITTF 2026) specify that a ball dropped from 30 cm onto a regulation table must rebound to approximately 23 cm. The corresponding laboratory value is:

$$CoR_{table} = \sqrt{\frac{23}{30}} = \sqrt{0.77} \approx 0.88 \tag{2.29}$$

Under match conditions, the effective restitution is lower. The ball arrives with oblique incidence, horizontal velocity and spin, introducing additional deformation modes and frictional dissipation.

Using match-play data reported by Zhou (2022), a translational speed of 18.4 m/s was observed to reduce to approximately 12.64 m/s after table contact, giving:

$$CoR \approx \frac{12.64}{18.4} = 0.68 \tag{2.30}$$

This represents a reduction of approximately 22% relative to the laboratory value. Velocity dependence is supported empirically by Inaba et al. (2017):

$$CoR = -0.0096 \times v_y + 1.0418 \tag{2.31}$$

where v_y is the pre-impact vertical speed in km/h. Restitution decreases linearly with increasing impact velocity. Representative values are provided in Table 2.4.

2.13.3 Bat Restitution, Dwell Time and Spin Ratio

For an inverted-rubber bat, the normal *CoR* lies in the range 0.75–0.85 (Kawazoe and Suzuki 2010), although this effective value depends on blade construction, sponge hardness, structural stiffness and overall configuration. Restitution is therefore not a fixed material property but a system-level response of the composite system under impact (Renaud et al. 2019).

Effective bat restitution is influenced by several interacting factors:

- **Blade stiffness**: Higher structural stiffness reduces bending losses and increases the proportion of recoverable normal impulse.
- **Sponge compliance:** Efficient elastic recovery enhances restitution, whereas excessively soft sponges increase internal damping and reduce restitution (Tiefenbacher and Durey 1994, Miyazawa et al. 2020).
- **Impact location**: The bat is not homogeneous. Impacts near the sweet spot (centre of percussion) minimise vibrational losses and maximise rebound efficiency, whereas off-centre impacts excite structural modes that dissipate additional energy and reduce *CoR*.

These same deformation dynamics also determine dwell time (t_d). This coupling has direct consequences for the spin ratio $S = \omega \cdot R/v$.

- **High-speed configurations**: A higher *CoR* generates increased rebound velocity. Although angular velocity also rises due to the deformation under forceful contact, the proportional increase in v typically dominates, resulting in a lower spin ratio.

- **Spin-oriented configurations**: Systems with longer dwell time and slightly reduced restitution promote sustained tangential traction, producing a higher value of S.

Thus, equipment defines the attainable range of spin ratio through its structural properties, whilst skill selects the operating point by controlling how impulse is partitioned between normal and tangential components during impact.

2.13.4 Reactive Mechanism and Blocking

The preceding sections establish intrinsic restitution characteristics of the ball–bat system. These define what the structure will return under given impact conditions. The player, however, can modulate how the rebound responds.

A training board (Fig. 2.19) illustrates a simple reactive return mechanism: if the incoming ball possesses sufficient linear momentum ($m \cdot v$) and/or angular momentum ($I_m \cdot \omega$), a valid return requires no additional mechanical work. The board simply reacts according to its own stiffness and restitution characteristics.

Unlike a fixed rebound board, the player has two options:

- **Energy injection**: Active force generation (F_h, F_b) increases the total mechanical energy imparted to the ball, augmenting or exceeding the incoming momentum.
- **Energy redirection**: Applied normal force is deliberately attenuated; incoming momentum is utilised and redirected, altering trajectory and spin through reactive return.

Thus, a *reactive return* (R_r) suppresses the active (hand-generated) normal force F_h, allowing the bat's passive elasticity to dominate. The outgoing ball departs quickly but carries limited *feedable* momentum, pressuring the opponent through placement and timing rather than momentum.

Although often described as *passive*, the block stroke is a technically demanding *reactive technique* requiring early perception, anticipation and precise control.

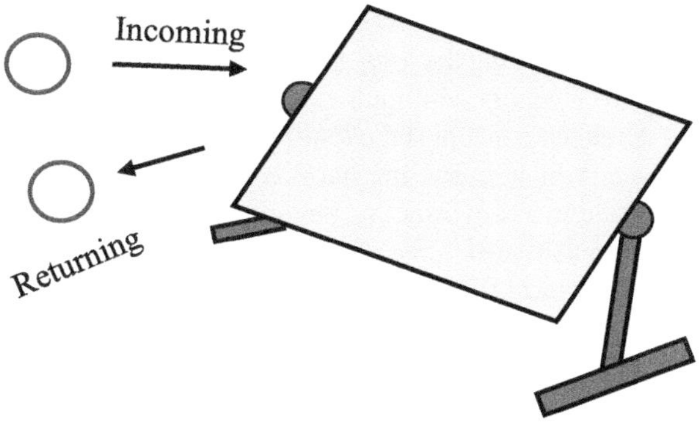

Fig. 2.19 Schematic of a reactive training board, illustrating a reactive return (R_r) system in which the return trajectory is governed primarily by the incoming ball's velocity and spin

Its mechanical complexity lies in managing substantial incident impulse whilst contributing minimal additional work.

Successful reactive control depends on several interrelated factors:

- **Anticipatory as a prerequisite**: Effective blocking begins before contact. Skilled players read the opponent's velocity, spin axis and intended contact region at an early stage, enabling timely body alignment and stabilisation of the bat prior to impact.
- **Bat angle as a primary determinant**: Fine adjustments in bat orientation, involving a slightly closed angle ($\beta < 90°$), govern rebound height, spin re-engagement and directional accuracy. Minute variations in angle can determine whether the block lands short or carries deep, reflecting the sensitivity of the incidence–exit motion.
- **Impulse shaping**: Even with minimal F_h, the player modulates contact through micro-adjustments of bat position and grip tension, cushioning, firming or stabilising the interaction. These refinements alter effective stiffness and damping during impact, producing outcomes ranging from an absorbing, energy-dissipative block to a compact punch-style counter. Through hand and wrist action, the player regulates the effective compliance and restitution of the ball–bat system, demonstrating how skill shapes the realised performance of the equipment (Sect. 2.2.2).
- **Proactive disruption**: A compact, well-timed block redirects the opponent's own momentum at uncomfortable angles or unexpected speeds. The stroke does not just return the ball; it converts incoming energy into renewed pressure on the opponent.

The block is *reactive* in its application of force, but *active* in perception and tactical intention.

2.13.5 Active and Hybrid Returns

The definition of a reactive return (R_r) as a passive block invites its natural counterpart: the *active return* (A_r). The active return encompasses the foundational offensive strokes, including the *loop*, *drive*, and *smash*, in which the outgoing ball is dominated by player-generated impulse.

A related concept is the *low-*F_h ball, where the normal hitting force is deliberately reduced whilst preserving stability and directional placement. This technique extends the reactive return principle beyond blocking to strokes such as slicing, chopping and certain push variations. Competitive play, however, rarely operates at the poles of passivity and full activity. Between the *dead block* and the *full swing* lies a crucial tactical space: *hybrid returns* – strokes such as the counter-hit, punch block and aggressive flick, where the player combines incoming momentum with a short, sharp impulse of their own.

Table 2.13 summarises the spectrum of return mechanisms.

Table 2.13 Spectrum of return mechanisms

Spectrum	Example	Force origin	Primary tactical intent
Reactive return (R_r)	Blocking	>90% Incoming ball	Redirection, survival or proactively disruptive
Hybrid return	Counter-hit, Punch	70% Ball, 30% Player	Disruptive
Active return (A_r)	Drive, Loop	>90% Player generated	Winning the point, temporal pressure

Table 2.14 Phenomenological mapping of energetic states

Qualitative descriptor	Physical state	Mechanical basis
Explosive	High E_v, high P	Rapid transfer of translational energy during a short contact interval
Loaded	High E_ω	Concentrated rotational energy; elevated torque demand upon impact
Floaty (hanging)	High E_ω, low E_v	Magnus-induced lift significantly influences trajectory relative to forward velocity
Dead (stopping)	Low E	Minimal total energy; 'reluctant' post-bounce motion

2.14 Energy and Power

Having established the force–impulse framework governing impact (Sects. 2.5 and 2.6) and the roles of friction, spin generation and transformation, and restitution (Sects. 2.8–2.13), the analysis now turns to the ball's kinetic energy, and the rate at which this energy is generated during impact.

2.14.1 Energy of the Ball

When a player applies *force* (F) during the brief contact interval, an *impulse (J)* is delivered, producing a change in *momentum* (p). F, J and p are vector quantities and do not describe the ball's capacity as a collective quantity to influence the subsequent exchange. That capacity is more completely represented by *kinetic energy*, the *disruptive potential* of the ball acting on the opponent.

The total kinetic energy of the ball (E) is the sum of its translational and rotational components:

$$E_v = \frac{1}{2}mv^2 \tag{2.32}$$

$$E_\omega = \frac{1}{2}I_m\omega^2 \tag{2.33}$$

where m is the ball's mass, v its translational speed, ω its angular velocity and I_m its moment of inertia.

Because energy scales *quadratically* with velocity ($E_v \propto v^2$ and $E_\omega \propto \omega^2$), modest variations in velocity can produce disproportionately large changes in the ball's energetic state, $E(v, \omega)$.

- **Higher-energy strokes** such as drives, topspins, smashes, forceful pushes and heavy chops impart substantial translational and/or rotational energy, producing trajectories that arrive with high velocity or aggressive 'jump'. These *high-E* balls carry both immediate threat and time pressure.
- **Lower-energy strokes**, including soft pushes, touch blocks, and guided returns, introduce minimal energy, producing a 'reluctant' motion and dead bounce that may be unfamiliar and disruptive.

The Energetic Nature of the Chop

The chop generates significant rotational energy ($E_\omega = \frac{1}{2}I_m\omega^2$) whilst the translational speed (v) remains modest. The resulting ball is energetically concentrated ('loaded', 'heavy'): difficult to counter, awkward to time and tactically disruptive.

2.14.2 Power: The Rate of Energy Production

Power (P) is the rate at which that energy is transferred during impact. It reflects not only the magnitude of energy imparted but also the speed at which the transfer occurs.

The average power delivered to the ball during impact is expressed as

$$P = \frac{\Delta E}{\Delta t} = \frac{\Delta E}{t_d} \tag{2.34}$$

where ΔE represents the change in the ball's total kinetic energy over the duration of contact (t_d).

Because energy scales *quadratically* with velocity, small variations in bat velocity at separation produce disproportionately large changes in power. The decisive factor is not sustained velocity alone, but the acceleration achieved during the brief impulsive phase. A compact stroke with high terminal acceleration over a short interval yields greater average power than a longer, slower swing that achieves the same contact duration but a lower velocity increment.

2.15 Summary: Impact Mechanics

The physics of impact determines everything that follows. This chapter establishes the mechanical foundation linking bat configuration, ball states and stroke execution from which the trajectory dynamics of Chap. 3 and the tactical framework of Chap. 4 are subsequently built.

1. **The Equipment Paradox**

Modern sponge equipment combines *elastic restitution*, *structural stiffness* and *high surface friction*. This configuration enables:

- Strong rebound through the catapult effect
- Compact stroke production through brushing, poking and punching.
- Efficient spin generation through surface tribological traction and internal rubber hysteresis.

Consequently, powerful strokes can be produced with short, economical wrist movements.

At the opposite end, low-restitution bats favour control. Pips-out surfaces deliberately suppress frictional coupling, rendering the system spin-insensitive and allowing players to return the ball as though incoming rotation were largely absent.

This contrast constitutes the equipment paradox:

- High-performance bats expand attacking potential but demand refined control.
- Simpler bats sacrifice power to gain stability and placement precision.

Equipment choice, therefore, is not a preference but a commitment to a particular mechanical regime. This paradox also anticipates the broader mechanical evolution of the sport: from the control-dominated hardbat regime, through the elastic power of sponge, to the deliberate synthesis of contrasting surfaces on a single bat (Sect. 2.12.5).

2. **Ball States and Disruption**

Equipment and stroke action together determine the *ball states* delivered to the opponent. A mapping of ball energetic states to phenomenological observation is provided in Table 2.14. Two disruptive states are particularly important:

- **Low-F_h balls** carry limited *recoverable energy* and disturb stroke timing through deliberate reduction of forward impulse, arriving slowly, often with a floaty trajectory that unsettles rhythm.
- **Energy-depleted balls** carry little translational energy but retain substantial rotational energy (E_ω). These 'heavily loaded' balls impose a persistent test of the opponent's adaptability and patience.

Energy depletion may arise naturally from defensive rubbers but can equally be produced with inverted equipment through controlled attenuation of forward impulse (J_h), making disruption a choice, not a consequence of equipment.

3. **Stroke Quality and Technical Pressure**

The mechanical characterisation of a stroke is defined by its impulse and spin ratio:

$$\varphi = (J, S) \tag{2.35}$$

where

- $\boldsymbol{J} = (J_h, J_b)$: the impulse, normal J_h and tangential J_b components
- $\boldsymbol{S} = \omega \cdot R/v$: the spin ratio, describing the partition of velocity between rotation and translation

  They define the stroke's mechanical signature at impact.

Stroke Quality: Intrinsic

(J, S) characterises execution at contact. It is opponent-independent and does not yet imply tactical effect.

Technical Pressure: Relational

Technical pressure arises only when a given (J, S), interacts with an opponent. It reflects the processing demand imposed on their perception, anticipation and motor response. The same stroke may therefore produce high pressure against an unprepared opponent, yet little or none against an expert who anticipates it.

$$\text{Technical Pressure, } \varphi_{\text{opp}} = f((J,\ S), \text{Opponent Model}) \tag{2.36}$$

where the opponent model includes reaction window (t_w), rhythm and biomechanical limits.

Causal Chain

(J, S) at impact I_1 determines the ball state (v, ω) through flight and table contact, arriving at the opponent as a realised demand φ_{opp} at I_2. Technical pressure is therefore not a property of the ball alone, but of the opponent's difficulty in returning it.

Interpretation

Stroke quality defines potential threat; technical pressure is that threat realised. Thus, (J, S) must be adapted to the opponent rather than optimised in isolation.

References

Gossard T (2026) Learning racket-ball bounce dynamics across diverse rubbers for robotic table tennis. https://arxiv.org/abs/2604.11349

Heaton J (2016) Table tennis: skills, techniques and tactics. The Crowood Press, Ramsbury. ISBN: 978-1-84797-090-9

Hibbeler RC (2018) Mechanics of Materials. pp762 (cantilever beam bending), pp813 (Parallel-Axis Theorem). 10th Edition. ISBN: 978-1-292-17820-2

Inaba Y et al (2017) Effect of changing table tennis ball material from celluloid to plastic on the post-collision ball trajectory. J Hum Kinet 55:29–38. https://johk.pl/wp-content/uploads/2023/03/10.1515_hukin-2017-0004.pdf

International Table Tennis Federation (2026) Statutes. https://documents.ittf.sport/sites/default/files/public/2026-02/2026_Statutes_v1_consolidated_clean.pdf

Kawazoe Y and Suzuki D (2010) Prediction of table tennis racket restitution performance based on the impact analysis. Theor Appl Mech Jpn 52:163–174. https://www.jstage.jst.go.jp/article/nctam/52/0/52_0_163/_article

Leissa AW (1969) Vibration of plates. National Aeronautics and Space Administration, U.S.A. https://ntrs.nasa.gov/api/citations/19700009156/downloads/19700009156.pdf

Lionel M et al (2014) Vibro-acoustic of table tennis rackets: influence of plywood design parameters—experimental and sensory analyses. Procedia Eng 72:374–379. https://doi.org/10.1016/j.proeng.2014.06.068

Mina di Sospiro G (2015) The metaphysics of ping-pong. Quest Books, Wheaton, IL. ISBN: 978-0-8356-0942-5

Miyazawa Y et al (2020) Effects of pimple height of a table tennis rubber on ball rebound behaviour. Proceedings 49(1):55. https://doi.org/10.3390/proceedings2020049055.

Mohammed H et al (2019) A mathematical model for the table tennis Ma Lin ghost serve. Proceedings of the World Congress on Engineering. https://www.iaeng.org/publication/WCE2019/WCE2019_pp36-41.pdf

Nakashima A et al (2009) Modelling of rebound phenomenon between ball and racket rubber with spinning effect. 2009 ICCAS-SICE, Fukuoka, Japan, pp. 2295–2300. https://ieeexplore.ieee.org/document/5335158

Renaud G et al (2019) Tabletennis ball impacting racket polymeric coatings: Experiments and modelling ofkey performance metrics. Appl Sci 9(1):158. https://doi.org/10.3390/app9010158

Rinaldi RG et al (2016) Non-linearity of the ball/rubber impact in table tennis: experiments and modelling. Procedia Eng 147:348–353. https://doi.org/10.1016/j.proeng.2016.06.307

Russell DA (2018) Acoustics of ping-pong: vibroacoustic analysis of table tennis rackets and balls. J Sports Sci 36(23):2644–2652. https://doi.org/10.1080/02640414.2018.1462578

Tiefenbacher K and Durey A (1994) The impact of the table tennis ball on the racket (backside coverings). Int J Table Tennis Sci 2. https://sasportssience.blob.core.windows.net/ijtts/IJTTS_2_pdf%20files/IJTTS_2_1_14_Tiefenbacher_The%20Impact.pdf

Varenberg M and Varenberg A (2012) Table tennis rubber: tribological characterization. Tribol Lett 47:51–56. https://doi.org/10.1007/s11249-012-9961-4

Wu Y et al (2021) Table tennis: effect of humidity on racket rubber tribology. Tribol Lett 69:31. https://doi.org/10.1007/s11249-021-01404-2

Zhou X (2022) Explanation and verification of the rules of attack in table tennis tactics. BMC Sports Sci Med Rehabil 14:6. https://doi.org/10.1186/s13102-022-00396-3

Chapter 3
Trajectory Dynamics: Flight, Drag and Magnus Effect

Whilst the mechanics of impact define the generation of force, friction and velocity, the ball's subsequent motion is governed by an entirely different family of physical processes.

Once separation from the bat occurs, the ball enters a dynamic environment in which its velocity and orientation are shaped by gravity, aerodynamic drag, and the Magnus force, often with greater magnitude and subtlety than intuition suggests.

Quantifying these forces allows their magnitude to be mapped precisely, directly linking aerodynamic behaviour to timing, stroke selection and the tactical demands placed on the opponent.

3.1 Three Forces: Gravity, Air Drag and Magnus Forces

3.1.1 Gravity

Gravity is the constant and inescapable force acting upon the ball. For a standard 2.7 g ball ($m = 2.7 \times 10^{-3}$ kg), the gravitational force (weight, W) is:

$$W = mg = 0.0027 \text{ kg} \times 9.81 \text{ m/s}^2 \approx 0.0265 \tag{3.1}$$

This force is intuitively familiar through the simple act of holding the ball. It establishes the baseline against which aerodynamic forces produce measurable deviations in flight. Because the ball's weight is modest, drag and Magnus forces may reach significant fractions of W, thereby acquiring disproportionate influence over trajectory and tactical outcome.

X. Ji, *Spin, Speed and Strategy*, Mathematical Engineering,
https://doi.org/10.1007/978-3-032-25518-1_3

3.1.2 Aerodynamic Drag

Aerodynamic drag F_d acts opposite to the ball's instantaneous velocity vector v and is given by (Marie-Magdeleine 2024)

$$F_d = -\frac{1}{2}\rho C_D A v^2 \tag{3.2}$$

where:

- ρ is the air density ($\approx$ 1.225 kg/m^3),
- v is the ball's translational velocity,
- A is the cross-sectional area of the ball ($A = \pi R^2$), and
- C_D is the dimensionless drag coefficient (Table 3.1).

The quadratic dependence on velocity (v^2) implies that drag increases rapidly with speed, producing substantial deceleration over the short flight distances typical of table tennis. The dependence on cross-sectional area (R^2) is equally consequential. When the official ball diameter increased from 38 to 40 mm in 2000, the associated increase in A raised aerodynamic resistance, contributing to reduced speed and extended rally duration.

3.1.3 Reynolds Number and Aerodynamic Regime

The Reynolds number (Re) characterises the flow regime around the ball and determines the effective values of C_D and C_L.

$$Re = \frac{2R}{\vartheta}v = 2739v \tag{3.3}$$

where ϑ is the kinematic viscosity of air ($\approx$ 1.46 × 10^{-5} m^2/s), and v is the ball's translational velocity (Table 3.2).

Typical playing speeds of 3–30 m/s correspond to $Re \approx 8 \times 10^3$ to 8×10^4. Within this range, the ball remains in the *subcritical regime*, well below the 'drag crisis' transition $Re \approx (2.5 \times 10^5)$ identified by Achenbach (1974).

As a result, C_D and C_L do not drop abruptly but vary continuously with surface roughness, translational speed and spin ratio (S) (Schneider et al. 2022). For context, streamlined vehicles exhibit $C_D \approx$ 0.25–0.35 and a flat plate normal to the flow has $C_D \approx$ 1.28, illustrating how strongly C_D depends on geometry.

3.1.4 Aerodynamic Coefficients and Magnus Force

Both drag and lift coefficients display non-linear behaviour across the playing speed range (Schneider et al. 2022 and Ou et al. 2011), and neither can be treated in isolation from the other.

C_D decreases slightly with increasing speed for non-spinning balls but rises with higher spin ratios (Schneider et al. 2022). Representative values are given in Table 3.1.

Aerodynamic lift, the Magnus force F_m, is generated by ball rotation ω and expressed as (Marie-Magdeleine 2024):

$$F_m = \frac{1}{2}\rho A C_L v^2 = \frac{1}{2}\rho A C_m \omega v \tag{3.4}$$

where C_L is the dimensionless lift (Magnus) coefficient, and $C_m = C_L/S$. The direction F_m depends on the spin axis (Fig. 3.1):

- Downwards for topspin (σ)
- Upwards for backspin (∂)
- Lateral for sidespin (θ).

Over the Reynolds-number range relevant to table tennis, the lift coefficient is approximately proportional to the spin ratio $S = \omega \cdot R/v$, although this linearity is bounded.

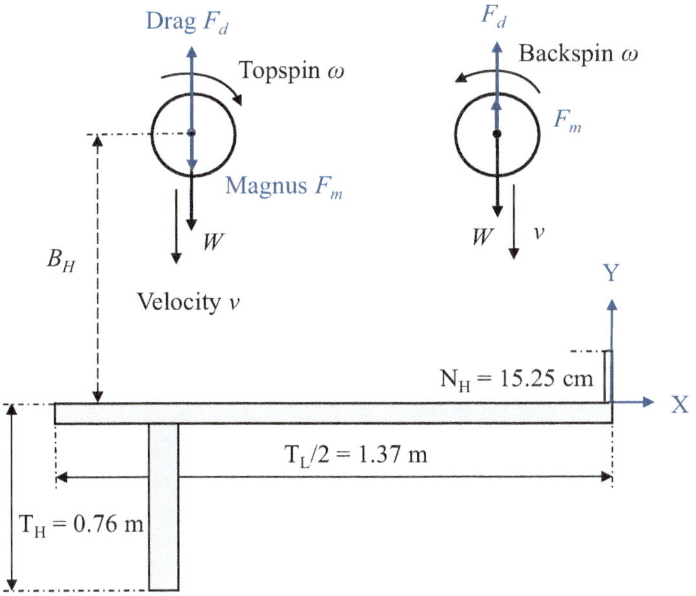

Fig. 3.1 Forces on a vertically descending spinning ball. Diagram illustrating the interaction of gravity, aerodynamic drag and the Magnus forces for topspin and backspin

Table 3.1 Representative ranges of drag coefficient (Nguyen et al 2019 and Schneider et al. 2022)

Play scenario	Velocity range (m/s), v	C_D
Slow rallies	2–8	0.58–0.63
Controlled topspin/block	8–14	0.52–0.58
Fast topspin/smash	14–20$^+$	0.42–0.52

Table 3.2 Magnus coefficients (Nguyen et al. 2019 and Schneider et al. 2022)

Spin ratio $S = \omega \cdot R/v$	C_L at lower Re	C_L at higher Re
Low spin, $S < 0.3$	-0.05–0.05	0.0–0.08
Moderate spin, $0.3 < S < 0.7$	0.10–0.22	0.18–0.28
Strong spin, $0.7 < S < 1.2$	0.22 –0.35	0.30–0.42
Extreme spin, $S > 1.2$	0.35–0.42	0.38–0.50

At very low spin ratios ($S < 0.3$), C_L stagnates or may even turn negative due to the inverse Magnus effect (Nguyen et al 2019). At high-velocity smashes, C_L values can exceed 0.5, significantly amplifying aerodynamic lift through the rotation–translation coupling (Ou et al. 2011). C_L values across the full range of S are given in Table 3.2.

The representative ranges in Tables 3.1 and 3.2 demonstrate that speed, spin and aerodynamic response evolve interdependently, reinforcing that trajectory prediction requires simultaneous consideration of translational and rotational motion.

3.2 Vertical Fall: Effects of Drag and Magnus

The lob is a tactical retreat from the table, temporarily relinquishing initiative in exchange for recovery time or positional reset. After reaching its apex, the ball enters a phase of near-vertical descent. The lob therefore provides the clearest practical example of predominantly vertical motion in table tennis.

During the descent phase (Fig. 3.1), the ball is governed by three principal forces:

- Gravity (W) is constant in magnitude and acting downwards.
- Aerodynamic drag (F_d) opposes the velocity and therefore acts upwards during descent.
- Magnus Force (F_m) acts downwards for topspin (σ) and upwards for backspin (∂).

Consequently, topspin lobs descend more steeply and rapidly, whereas backspin lobs exhibit prolonged hang time and greater sensitivity to speed loss and external (air movement) disturbances.

1. **Ideal parabolic trajectory (neglecting aerodynamics)**

If air resistance and spin are neglected, the motion is governed solely by gravity. For a ball descending from maximum height B_H to the table surface (with the table surface taken as the zero-height reference), the time t and velocity v of fall follow the standard kinematic relations:

$$v = gt \tag{3.5}$$

$$t = \sqrt{\frac{2B_H}{g}} \tag{3.6}$$

For a representative lob reaching $B_H = 4.5$ m:

$t \approx$ 0.96 s, $v \approx$ 9.4 m/s.

If both the ascending and descending flights (F_1 and F_2) are approximated as parabolic, the total reaction window available to the receiver is

$$t_w < t_{F_1} + t_{F_2} \approx 2t + 2t \times CoR = 2t(1 + 0.88) = 3.61s$$

This simplified estimate provides a quantitative baseline: a high lob affords less than four seconds of total response time, establishing a lower bound for perceptual and motor preparation.

2. **Including drag and spin: equation of motion**

When drag and spin are included, the vertical equation of motion (taking downward velocity as positive) becomes

$$m\frac{dv}{dt} = mg \pm F_m - F_d \tag{3.7}$$

where the sign of F_m depends on the spin direction (positive for topspin σ, negative for backspin ∂).

Grouping the coefficients yields the governing differential equation:

$$\frac{dv}{dt} = g - bv^2 \tag{3.8}$$

where b is an effective aerodynamic parameter incorporating both drag and Magnus contributions:

$$b = \frac{1}{2m}\rho A(C_D \mp C_L) \tag{3.9}$$

Assuming a moderate spin ratio $S = 0.3$, representative coefficients are $C_D \approx 0.55$ (Table 3.1) and $C_L \approx 0.3$ (Table 3.2). Using $R = 20$ mm and $m = 2.7$ g, the resulting parameters are:

$b_\sigma = 0.072$ (topspin), $b_\partial = 0.243$ (backspin)

Simulated results for $B_H = 4.5$ m

Numerical solutions of Eqs. (3.8–3.9) demonstrate the pronounced influence of spin as shown in Fig. 3.2.

As shown in Table 3.3, topspin increases the effective downward acceleration through the combined action of gravity and Magnus force, reducing flight time and increasing impact velocity. Backspin acts in opposition, reducing descent rate and prolonging flight.

Flight duration is therefore directly tuneable through spin, confirming spin as a temporal control variable in addition to its spatial and rotational effects.

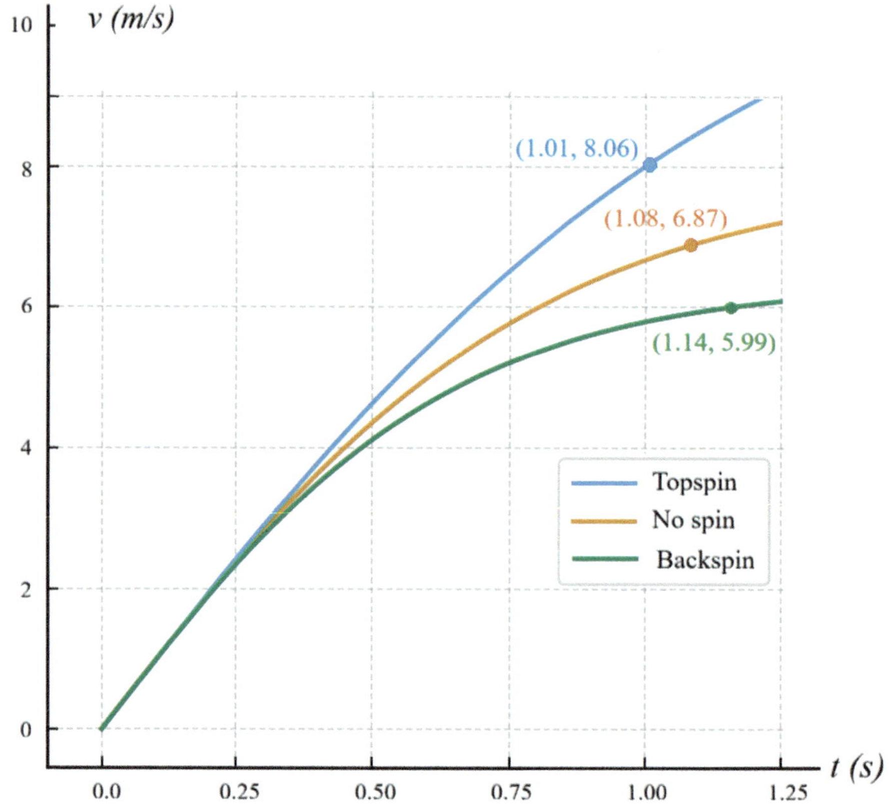

Fig. 3.2 Velocity increase over time for a falling ball with topspin, no spin and backspin. The graph shows that topspin accelerates descent through a downward Magnus force, whilst backspin provides lift and produces the slowest velocity increase, thereby altering the player's reaction window. The plotted data are from the author's 2023 study on table tennis physics

Table 3.3 Time of fall and impact velocity

Spin condition	Time of fall	Impact velocity	Effect versus No-spin
Topspin (σ, $C_D + C_L$)	1.01 s	8.06 m/s	Shorter time ($\approx$ 6.5% faster), higher v
No-spin (χ, C_D)	**1.08 s**	**6.87 m/s**	**Baseline aerodynamic deceleration**
Backspin (∂, C_D-C_L)	1.14 s	5.99 m/s	Longer time ($\approx$ 5.5% slower), lower v

3.3 Flight Time and Trajectory

The previous section examined the ascending and descending motion of the ball. The present section extends the analysis to horizontal motion and more realistic playing scenarios. As before, the discussion begins with an idealised case neglecting air resistance and spin and then introducing aerodynamic drag and Magnus effects.

Where full aerodynamics are considered, reference is made to the numerical simulations of Ou et al. (2011) which solve the non-linear equations of motion whilst incorporating the dependence of drag and lift coefficients on velocity and spin. These results provide a benchmark against which simplified analytical models may be assessed.

3.3.1 Baseline Flight without Drag or Spin

Consider a shot played from one baseline to the other, just clearing the net (Fig. 3.3).

Neglecting air resistance and spin, the ball follows a parabolic trajectory governed by classical projectile motion. Let x denote the horizontal displacement and y the vertical displacement:

$$x = v_0 \cdot cos\alpha \cdot t \tag{3.10}$$

$$y = v_0 \cdot sin\alpha \cdot t - \frac{1}{2}gt^2 \tag{3.11}$$

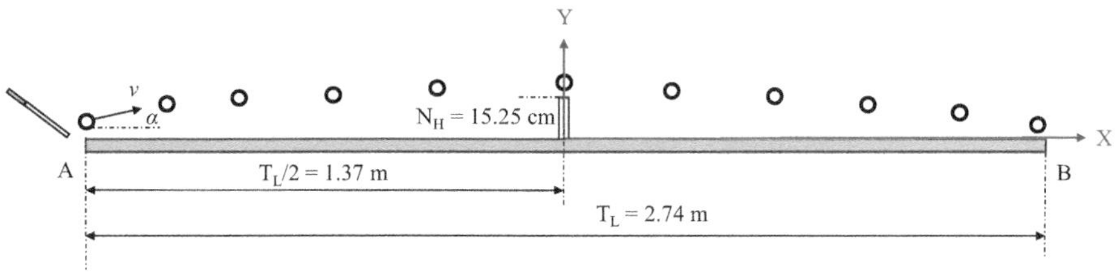

Fig. 3.3 Two-dimensional kinematic diagram of ball flight. The trajectory is shown from impact point at (A) to table landing at point (B), with launch angle α, net height 15.25 cm and table length 2.74 m indicated

Eliminating t gives the trajectory equation:

$$y(x) = xtan\alpha - \frac{gx^2}{2v_0^2cos^2\alpha} \tag{3.12}$$

For a *just-clearing* shot launched from the table height ($y_0 = 0$), two boundary conditions apply:

1. **Landing at the far baseline**

   $y(\mathrm{T_L}) = 0$, where $\mathrm{T_L} = 2.74$.
   Substituting into Eq. (3.12) yields the range condition:

$$v_0^2 = \frac{g \times \mathrm{T_L}}{\sin(2\alpha)} \tag{3.13}$$

2. **Clearing the net at mid-table**

   $y(\mathrm{T_L}/2) = \mathrm{N_H}$, where $\mathrm{N_H} = 15.25$ cm.
   This constraint gives the required launch angle:

$$\alpha = \arctan\left(\frac{4\mathrm{N_H}}{\mathrm{T_L}}\right) \tag{3.14}$$

The total flight time for a projectile returning to its launch height is

$$t_f = \frac{2v_0sin\alpha}{g} \tag{3.15}$$

Substituting Eq. (3.13) into Eq. (3.15) gives

$$t_f = \sqrt{\frac{2\mathrm{T_L}}{g}tan\alpha} \tag{3.16}$$

Using standard table dimensions ($\mathrm{T_L} = 2.74$ m, $\mathrm{N_H} = 0.1525$ m, g = 9.81 m/s^2):

- **Launch angle**: $\alpha = \arctan(4 \times 0.1525/2.74) \approx 12.5°$
- **Launch speed**: $v = \sqrt{(9.81 \times 2.74)/\sin(25°)} \approx 8.0$ m/s
- **Flight time**: $t_f = \sqrt{2 \times 2.74\tan(12.54^{\circ})/9.81} \approx 0.35$ s.

This idealised case establishes a quantitative benchmark: a full-table shot must be launched at approximately 8 m/s and 12.5° to reach the opponent's baseline. The resulting flight time of approximately 0.35 s illustrates the narrow reaction window (t_w) associated with a typical attacking stroke.

3.3.2 Real Flight with Drag and Magnus

When aerodynamic drag and Magnus forces are included, the ideal parabolic trajectory becomes distorted. The computational study by Ou et al. (2011) provides representative solutions obtained through numerical integration of the full non-linear equations of motion, revealing the complementary aerodynamic behaviour that precedes and follows the table contact and the associated transformation of spin ratio.

Figure 3.4 shows simulated trajectories for a launch speed of 10 m/s, a launch angle of 10° and a spin ratio ($S = \omega \cdot R/v = 0.5$), consistent with the analytical trends predicted by Eq. (2.23) and the results shown in Fig. 3.2.

Key behaviours:

- **Backspin (ə)**: Upward Magnus force partially offsets gravity, increasing arc height and flight time.
- **Topspin (σ)**: Downward Magnus force reinforces gravity, producing a shorter, dipping trajectory.
- **Sidespin (θ)**: Introduces lateral curvature, with limited effect on flight time or peak height unless coupled with topspin or backspin.

These trajectories reflect a division of physical roles: drag governs speed decay, the Magnus force governs curvature, and gravity sets the vertical reference. Tactically,

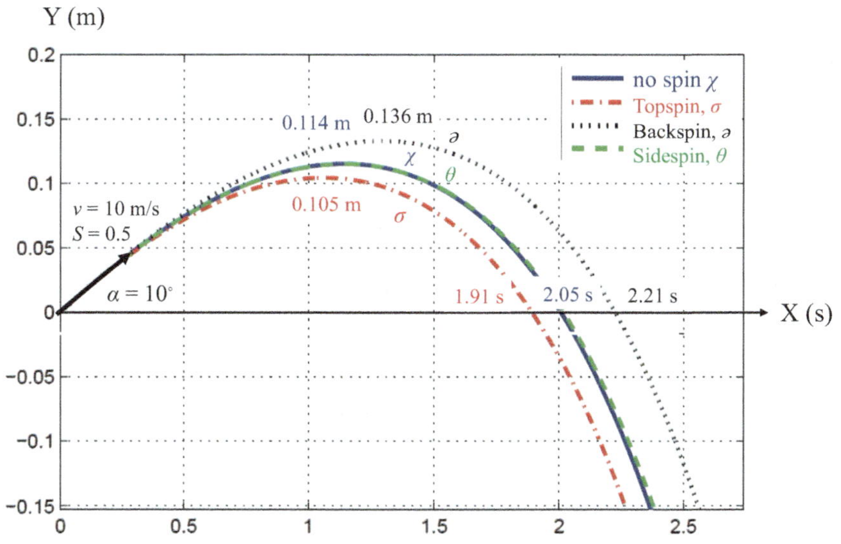

Fig. 3.4 Simulated vertical position versus horizontal distance for a table tennis ball under different spin conditions. Backspin, no-spin and topspin trajectories are compared for a ball launched at 10° and 10 m/s with spin ratio of 0.5. Topspin shortens flight duration and reaction interval, whilst backspin extends them. After Ou et al. (2011), Fig. 16a. Reproduced with permission from Dr Kui Ou

this translates into time manipulation: topspin compresses, backspin extends and sidespin diverts displacement.

The inclusion of this full non-linear solution supports the book's core analytical commitment: a physically consistent treatment of the ball across flight, bounce and impact, validated by independent computational methods.

3.3.3 *Relative Magnitudes of the Forces*

High-speed footage from elite competition, including the 2024 Paris Olympics, shows topspin strokes exceeding 120 rps. For a representative stroke:

$v = 18.4$ m/s, $\omega = 2\pi \times 120 = 754$ rad/s,

giving a strong spin ratio.

$S = \omega \cdot R/v \approx 0.82$.

with a corresponding Magnus coefficient $C_L \approx 0.8$ (Table 3.2).

The resulting force magnitudes are approximately (Fig. 3.5):

- Gravity $W \approx 0.0265$ N
- Drag $F_d \approx 0.12$ N (for $C_D \approx 0.5$)
- Magus force $F_m \approx 0.016$ N.

Drag is therefore the dominant force at high speed, driving rapid deceleration. The Magnus force shapes the trajectory through curvature, whilst gravity provides a constant vertical bias. The observed flight path arises from their continuous vector interaction.

3.4 Linking Flight Dynamics to Gameplay Strategy

Ball trajectory is not a passive background to the rally; it is information-rich and continuously shapes, limits and enables tactical choice. Translational speed, spin-induced curvature and drag-governed deceleration together determine *when* interception is possible and *which* strokes are viable at that instant.

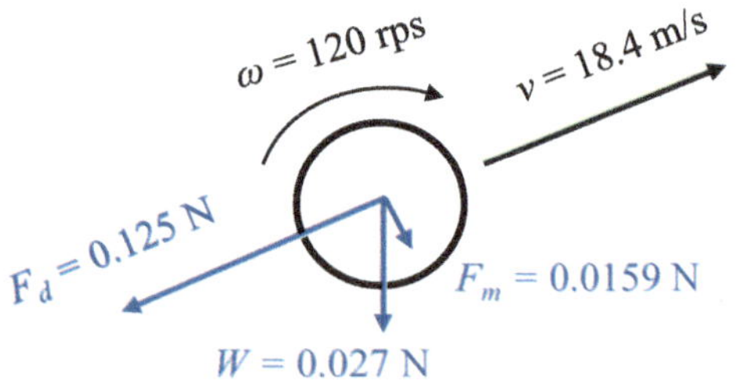

Fig. 3.5 Force magnitudes acting on a table tennis ball travelling at 18.4 m/s with a spin rate of 120 rps, showing weight (0.027 N), aerodynamic drag (0.125 N) and Magnus force (0.0159 N)

3.4.1 Timing as Aerodynamic Selection

Timing depends on the player's perception of ball position and velocity during flight (F_1 and F_2). The chosen interception point along the post-bounce trajectory (F_2) is strategic rather than inevitable and is formally defined as *interception timing* (τ). It determines both the moment of contact, and the bounce phase engaged.

During flight, gravity, Magnus force and aerodynamic drag act continuously. Drag dominates translational decay: with $F_d \propto v^2$, translational speed and kinetic energy (E_v) diminish rapidly, whereas angular velocity ω is comparatively preserved, increasing the spin ratio (S). Interception timing is therefore the deliberate selection of this evolving state. Three principal phases can be distinguished.

- Early interception (ascending phase, τ_{rising}, near table contact TC)

Contact occurs before drag has substantially reduced velocity, enabling flatter trajectories, rapid redirection and maximal compression of the opponent's response window. This phase prioritises initiative and temporal pressure.

- Peak interception (τ_{peak}, apex of F_2)

Vertical velocity approaches zero, offering geometric stability. Speed has moderated whilst spin influence has increased. This phase favours precision and placement-led strategy.

- Late interception (τ_{falling}, descending phase of F_2)

Extended flight amplifies the filtering effect. Both speed and spin decay further. The reduced pace permits more composed execution but concedes table position and increases vulnerability to short-ball exploitation.

The optimal interception phase is therefore not *as early as possible* but as early as composure permits. With response windows often below 0.35 s, players rely on anticipation, micro-adjustments in footwork and pattern recognition to align interception timing with tactical intent-pressure (τ_{rising}), precision (τ_{peak}) or control (τ_{falling}).

Thus, drag functions as a temporal selector: by choosing when to intervene, the player selects which physical regime of the ball to confront.

3.4.2 Spin: Disruption and Its Counters

If speed represents the thesis of attack in table tennis, an assertion of force, then spin is its antithesis, a disruptive mechanism that reshapes flight, modifies bounce behaviour and challenges perceptual expectation.

- **Backspin** generates aerodynamic lift, producing slower, floating trajectories that delay timing and invite misjudgement.
- **Topspin** induces downward Magnus acceleration, creating pronounced dip and forward kick that rewards early interception.

- **Sidespin** introduces lateral curvature and spatial ambiguity, complicating prediction.

Spin, however, is not without cost. Generating spin (Sect. 2.11.1) demands deliberate tangential brushing, a more intentional action than the instinctive explosive strike, and one that requires additional preparation time. It also demands sufficient vertical (or lateral) clearance to drag the bat across the ball's surface: upwards for topspin, downwards for backspin, and laterally for sidespin. Both constraints reduce forward impulse (J_h) and lengthen the stroke cycle. Skilled receivers exploit this by:

- Blocking into the spin, converting stored rotational energy (E_ω) through bat restitution (CoR).
- Redirecting extreme incoming spin with a soft under- or side-cut, using minimal added force so that the ball's own rotational energy is preserved and returned. The initiator must now contend with the heavy spin they generated – their mechanical advantage reversed against them.
- Applying flat counterattacks dominated by linear impulse to override curvature.
- Using damped or low-friction equipment (e.g. pips-out or anti-spin rubbers) to weaken tangential coupling and simplify the incident spin profile.

References

Achenbach E (1974) The effects of surface roughness on the heat transfer and flow past spheres. J Fluid Mech 65(1):113–125. https://www.cambridge.org/core/services/aop-cambridge-core/content/view/B9276280823D702358961E435CA661C8/S0022112074001285a.pdf/the-effects-of-surface-roughness-and-tunnel-blockage-on-the-flow-past-spheres.pdf

Ou K et al (2011) Computational sports aerodynamics of a moving sphere: simulating a ping-pong ball in free flight. In: AIAA 29th Applied Aerodynamics Conference. https://doi.org/10.2514/6.2011-3668

Schneider R, Lewerentz L, Kemnitz S, Schultz C, Abou Jaoudé A (2022) Simulation modeling - recent advances new perspectives and applications. In: Table tennis and physics. IntechOpen. https://doi.org/10.5772/intechopen.107026

Nguyen S, Corey M, Chan W, Greenhalgh ES, Graham JMR (2019) Experimental determination of the aerodynamic coefficients of spinning bodies. Aeronaut J 123:(1263)678–705. https://doi.org/10.1017/aer.2019.15

Marie-Magdeleine AS (2024) Fluid mechanics. Chapter 6: Flow over bodies. Toronto Academic Press. ISBN: 978-1-77956-196-1

Chapter 4
Strokes, Serves and Tactics

4.1 Introduction

Winning and losing are integral to table tennis. Victory consolidates confidence; defeat exposes structural weaknesses and refines judgement. Competitive mastery requires that technical execution and tactical choice conform to the actual mechanics of impact and flight.

The analytical vocabulary is now established: the player's normal and tangential forces (F_h, F_b) and their corresponding impulses (J_h, J_b), modulated by dwell time (t_d), determine the spin ratio $S = \omega \cdot R/v$, leaving the bat. In flight (F_1, F_2), aerodynamic drag and Magnus lift transform the trajectory, whilst at the table contact (TC) the outcome is further transformed by table restitution (CoR) and friction (μ).

Within this physically grounded framework, tactics acquire a precise mechanical basis. Reactive and active returns represent distinct regimes of impulse management. An energy-rich or energy-depleted ball determines the energetic burden imposed on the receiving player.

Initiative, introduced here as a bridge from mechanics to tactics, is not synonymous with aggressiveness. It denotes the capacity to adapt within mechanical and physical constraints. Force applied without recalibration dissipates initiative rather than sustaining it. Advantage is maintained through the adaptation of intent, transforming stroke quality (J, S) into technical pressure (φ_{opp}).

At critical scorelines (e.g. 9:9 or 8:10), choices such as serving short or long are not exploratory but inferential, drawing on accumulated information about response patterns, positional tendencies and tolerance to spin–speed conditions. Tactical decision-making is therefore governed by the interpretation of information under pressure.

These principles provide the analytical foundation for the strokes, serves and tactical configurations examined in this chapter, forming a transition from mechanics to competitive decision-making.

X. Ji, *Spin, Speed and Strategy*, Mathematical Engineering,
https://doi.org/10.1007/978-3-032-25518-1_4

4.2 Representative Strokes

The three strokes examined here, push, topspin and smash, are not selected for their frequency alone, but because they span the full energetic and impulse range of ball–bat interaction. Each is mechanically distinct: the push is defined by open-angle, tangential under-contact; the topspin by closed-angle brushing with precise rotational coupling; the smash by normal-force dominance with minimal spin. Yet beyond mechanics, each stroke exposes a different relationship between technical intent and executional fidelity, and a different set of cognitive and psychological demands on the player. The push rewards patience and deception; the topspin demands coordination under extreme time pressure; the smash tests impulse control precisely when urgency is greatest. Understanding how these strokes work, and why they break down, reveals a deeper structure that is formalised in Sect. 4.3.

4.2.1 The Push Stroke and Backspin

The push is the most frequently used stroke in table tennis and is often the first to be learned. Despite its apparent simplicity, it is technically demanding and tactically significant. As a predominantly defensive action, the push requires precise regulation of contact angle (β), timing (τ) and tangential brushing force (F_b).

Mechanically, the stroke is executed by contacting the lower rear surface of the ball with a downward tangential motion. Pushes are most effective when the ball is contacted close to the net and below net height (N_H; Fig. 4.1). The bat angle must therefore be open ($\beta > 90°$), the normal force (F_h) carefully moderated and brushing force (F_b) intentionally prominent.

1. **Practical Control Principles**

Length control

Length is governed primarily by the hitting impulse (J_h). A reduced J_h produces short placement. An intermediate J_h yields half-court depth. An increased J_h drives

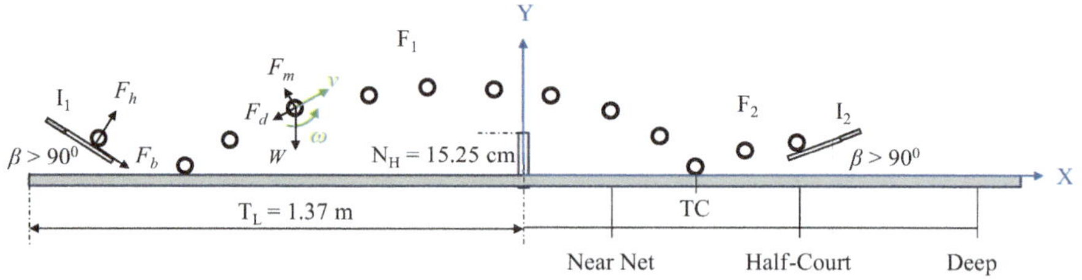

Fig. 4.1 X–Y plane schematic of stroke phases. The sequence shows initial impact (I_1), primary flight (F_1), table contact (TC), secondary flight (F_2) and opponent contact (I_2)

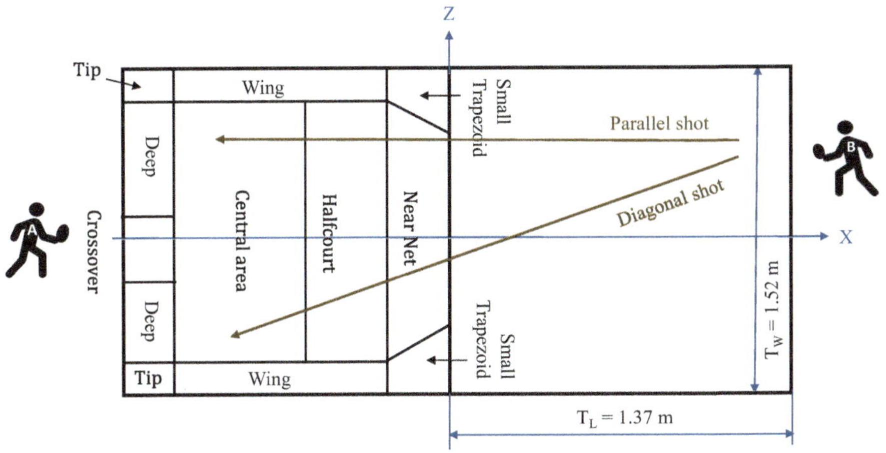

Fig. 4.2 Top-view schematic of twelve tactical zones. The table is divided into tactical regions, including Small Trapezoids, Near-Net, Hal-fcourt, Central, Wings and Deep zones, with the crossover point identified and parallel and diagonal trajectories distinguished

the ball deep. The margin for error is small. If J_h exceeds the required threshold, the ball travels beyond the intended zone (Fig. 4.2).

Spin and tempo variation

Spin magnitude depends on the tangential impulse (J_b). Increased J_b generates heavy backspin and may induce strong post-bounce deceleration or even velocity reversal (Sect. 2.11.3). If the receiver anticipates the wrong decay behaviour, contact timing shifts (τ), resulting in mistimed impact at I_2.

2. **The Mechanical Duality of the Push**

The push is often classified as an energy-depleting stroke. With low-F_h and significant tangential contact, translational speed is reduced, and backspin dominates post-bounce behaviour (Sects. 2.11.3 and 3.3.2). In this regime, the push stabilises tempo and delays attack.

However, the same mechanics allow a different outcome. A deep, fast push executed with decisive F_h and sustained tangential backspin contact generates a long trajectory. The ball reaches the opponent quickly and often forces a pressured lift.

Control of the push duality (short or long pushes) is to determine whether the stroke delays conflict or accelerates it.

3. **Counters to Backspin**

Effective response depends on recognising the incoming energy state, depleted or rich.

- **Stabilising**: The receiver accepts the energy-depleted regime, maintaining push exchanges and using controlled placement to wait for an opening.

- **Initiating topspin**: The receiver transitions to an energy-rich regime by applying a substantial brushing force (F_b) upwards to generate topspin. If F_b is insufficient, the incident momentum dominates, and the outgoing trajectory collapses into the net.

4.2.2 The Topspin Stroke

The topspin stroke (σ) defines the modern offensive game. Its effectiveness is enabled by the spring–mass–damper behaviour of contemporary blade–rubber systems, which amplify stored elastic energy through the catapult effect.

The stroke is executed by brushing upwards across the ball's *shoulder* with a closed bat angle ($\beta < 90°$), a movement that operates within a narrow tolerance band. Mechanically, the outgoing trajectory is a precise function of the bat's state, requiring the brushing impulse J_b, hitting impulse J_h and contact angle β to be perfectly synchronised to resolve the necessary velocity and spin vectors. In competition, this coordination must occur under extreme temporal pressure, with professionals required to identify spin and initiate a response in approximately 0.35 s (Sect. 3.3.1). Under such pressure, 'choking' manifests as a collapse in this high-speed internal timing rather than a failure of tactical intent. The inherent instability of the stroke reflects a reduction in executional fidelity, where the physical input fails to meet the rigid numerical requirements of the intended 'stroke effect'.

1. **Mechanics and Flight**

A topspin ball is characterised by a high spin ratio $S = \omega{\cdot}R/v$, which generates a downward Magnus force F_m. Spin ratio evolution is phase dependent. During flight (i.e. F_1 or F_2), differential aerodynamic decay increases S (Sect. 2.11.2). At table contact, frictional interaction decreases S (Sect. 2.11.3).

2. **The Block Response: Reactive, Precise and Dangerous**

Every dominant stroke invites its counter. The block is the primary close-table response to topspin (Sect. 2.13.4). The topspin player commits to a full rotational arc; the blocker compresses time.

The block does not resist incoming topspin; it appropriates it. By stabilising the bat early, reading the approaching trajectory and spin axis, and presenting a closed angle ($\beta < 90°$), the blocker shapes the outgoing impulse whilst contributing minimal hitting force (F_h).

This economy of action produces a distinctive tactical effect. A compact, well-timed block reverses the pace, returning a fast, low trajectory that compresses the topspin player's recovery window. Because the topspin stroke leaves the body biomechanically extended, weight forward, arm rotated, trunk open, the blocker's return arrives before the attacker can fully reset. The block thus exploits the temporal vulnerability inherent in the attacking stroke.

What makes the block dangerous is not its power but its placement and timing. The blocker does not win the point; the attacker loses it by being caught out of position. In this sense, the block operates at the reactive end of the return spectrum (Sect. 2.13.5), but its tactical intelligence, reading spin, anticipating direction, stabilising before impact, places it firmly within skilled, active perception.

3. **Alternative Counters**

When time and positioning allow, further responses become viable:

- **Counter-topspin**: Executed on the rise (τ_{rising}). Matches or exceeds the incident spin and counters topspin with a topspin.
- **Drive or flat hit**: Relying on strong J_h to override the approaching rotational curvature with dominant linear impulse.
- **Chop**: Typically, from mid-distance, reverse spin state, $\sigma \rightarrow \partial$.

Against topspin, counter-topspin, drive, flat hit or chop may all succeed; a push, however, rarely works, as the open bat angle tends to lift the ball high and surrender control. A well-timed block remains the most reliable and often the most dangerous counter to aggressive topspin.

4.2.3 The Smash Stroke

The smash is a decisive attacking stroke used against high or weak returns. It is executed by minimising or omitting the tangential brushing component, producing a powerful, flat hit (F_h>>F_b).

For beginners, smashing often feels instinctive and cathartic. True technique refines this instinct into a controlled, purposeful action. An effective smash requires anticipation, precise timing and intentional placement, converting a clear opportunity into a high-percentage winner rather than a hurried error.

The very urgency and explosiveness of the smash make it mechanically simple but operationally fragile. Under pressure, players may rush the motion, over-rotate or strike too early; small timing errors are sufficient to send the ball long or into the net. As with the 'choke' observed in the topspin stroke, the smash exposes the boundary between instinctive action and deliberate control. Precision depends as much on composure and impulse control as on biomechanics.

1. **The Strategic Role of the Smash**

The smash is a finishing mechanism. Its purpose is resolution. At advanced levels, instinct must be regulated. Acceleration without selection is unstable; a smash executed in frustration or haste sacrifices power, placement or timing, allowing the opponent to recover.

When impulse overrides control, the error is self-generated. An attack delivered without adaptation does not secure dominance; it relinquishes it. Properly executed, the smash is not just loud; it is final.

2. **Counters to the Smash-Managing Incoming Energy**

A smash delivers a sudden and dominant influx of translational kinetic energy (E_v) with minimal rotational energy (E_ω). The receiver must manage a large incoming normal impulse (J_h) within a severely reduced reaction interval (t_w).

- **Block**: A firm, stable bat angle with minimal added F_h. The technique is reactive return. Success depends on timing precision and bat angle discipline (Sect. 2.13.4).
- **Chop**: The defender (I_2) retreats to extend temporal margin, intercepting the falling phase (τ_{falling}). A *low-F_h, high-ω* action replaces incoming translational energy with rotational energy, shifting the rally to a E_ω-dominated state (Sect. 2.14.3).
- **Counter-smash**: Meets force with force. Demands balance and stable footwork to exploit the attacker's pace, relying on a small but decisive additional F_h to reverse direction and sustain depth.

Each counter represents a different energy-management strategy: reflect, dissipate or amplify.

4.3 The Stroke Spectrum

The push, topspin and smash span the primary energetic and impulse configurations of table tennis. Together, they reveal a deeper structure: stroke production reduces to a minimal set of contact primitives defined by *where* the ball is engaged and *how* impulse is oriented.

For the forehand and backhand, this reduces to three fundamental actions:

- **Lower-rear-quadrant contact** (Fig. 2.9b) directs the ball upward and forward over the net.
- **Upper-rear-quadrant contact** directs the ball downward and forward over the net.
- **Orthogonal component** introduces lateral brushing (left/right), generating sidespin and its compound forms.

These fundamental actions define the admissible contact space of ball–bat interaction, which may be organised as in Table 4.1.

All strokes, therefore, reduce to combinations of these primitives:

- **Under-contact, backspin-dominant class** (forehand and backhand): sway, push, slice, chop
- **Over-contact, topspin-dominant class** (forehand and backhand): flick, drive, slow topspin, fast topspin (loop)
- **Lateral (vertical axis) class**: lateral swipe and its compound forms

Table 4.1 Fundamental strokes

Stroke	Mechanical mandate	Stroke quality (J, S)	Tactical function
Push	Open-angle bat ($\beta > 90°$), under-contact, backspin	Low F_h, moderate S	Control, temporal extension
Slice (long push)	Open-angle bat ($\beta > 90°$), extended under-contact	Coupled $J_h + J_b$, backspin ∂, moderate–high S	Induce error, temporal extension
Topspin (loop)	Close-angle bat ($\beta < 90°$), over-contact with tangential engagement	Coupled $J_h + J_b$, topspin σ, moderate–high S	Pressure generation, initiative building
Smash	Close-angle bat ($\beta < 90°$), normal-force dominant	Dominant F_h, low S	Point termination, resolution

These classes arise not from *coaching* convention but from the intrinsic mechanics of ball–bat interaction. They constitute a mechanically complete basis for stroke production.

In practice, many players, even at advanced amateur level, exhibit incomplete coverage of this possibility space, with one or more classes underdeveloped or absent. Such limitations are *structural* rather than technical: missing a region of the contact space creates predictable vulnerabilities that can be systematically exploited.

Mechanical completeness, therefore, is not the ability to execute many variations of a few strokes, but the ability to access and control the full range of admissible interactions defined by under-contact, over-contact and lateral modulation. This same continuous space is later discretised and counted in Sect. 6.8, where the three contact primitives are expanded into a finite 'stroke space' that clarifies how mechanical completeness can be progressively trained and internalised.

4.4 The Serve

4.4.1 ITTF Service Rules

The serve is the only stroke initiated without incoming velocity, spin or spatial constraint. It therefore carries disproportionate tactical weight. A failed return of serve is among the most disheartening outcomes in table tennis, not just a lost point but a fracture in concentration that can echo through subsequent exchanges.

To maintain fairness between server and receiver, the International Table Tennis Federation (ITTF) prescribes strict mechanical and visibility requirements (ITTF 2026):

- Starting position: The ball must start behind the end line and above the playing surface.
- The toss: The ball must be projected near-vertically to a minimum height of 16 cm.

- Contact: The ball must be struck during its *downward* trajectory.
- The bounce: The ball must first contact the server's side before crossing the net.
- Visibility: The ball must remain continuously visible to the receiver and must not be obscured by the server's body, arm or clothing.

These regulations limit concealment and prevent smash-like service actions. The opening exchange is thereby preserved as a contest of spin generation, recognition and control rather than speed alone.

Technical Clarification

- **Unconstrained action**: Unlike rally strokes, the serve begins from a zero-state condition. The server specifies the initial values of spin (ω), speed (v) and placement (z) without interference. Here, z denotes the zone, corresponding to the regions shown in Fig. 4.2.
- **Near-vertical toss**: Defined as a projection with minimal angular deviation from the vertical (commonly interpreted as within 30°). This prevents body-assisted concealment of contact.

4.4.2 Serve Variation

The serve is the most information-dense action in table tennis. Unlike any other stroke, it lets the player engineer the rally's initial conditions, i.e., spin (ω), speed (v), placement (z) and tactical intent, all before the opponent touches the ball. This section examines how serve variation creates strategic control, and how the subsequent third ball contact (I_3) transforms that control into points.

1. **Placement and perceptual load**

Elite match analysis indicates a dominance of short forehand serves. Djokic et al. report that 76.9% of all analysed serves are forehand *short serves* (Djokic et al. 2013), designed to *bounce twice* on the receiver's side, within small-trapezoid zones near the net or the half-court (Fig. 4.2).

Short placement performs two structural functions:

- **Constraint**: Limited depth reduces the receiver's ability to execute high velocity attacking strokes.
- **Predictability induction**: By suppressing full-power attacks, the server narrows the range of possible returns.

The server's advantage is only as strong as the receiver's perceptual limitations. The ability to interpret spin is the primary differentiator between elite and amateur performance. Xu et al. (2019) demonstrate that professional players identify spin types with significantly higher accuracy than amateurs (78.4% vs 54.9%). Compound spin increases perceptual difficulty: approximately 25% of observed serves involve mixed components, most commonly side-backspin (θ_{∂}). Professionals also exhibit faster processing, with mean response times of approximately 0.55 s compared with 0.65 s for amateurs (Tian and Xiao 2024). A server who combines spin complexity with reliable short placement therefore compounds the receiver's task, forcing high-stakes judgement under severe time pressure.

2. **Long serves and temporal disruption**

Whilst short serves dominate statistically, selective long serves extend the second flight (F_2) onto the deep zones and wings. Their function is disruption:

- **Surprise** the opponent by disrupting their timing.
- **Draw weak returns** by catching the receiver out of position.
- **Initiate a *third-ball attack*** through forced passive returns.

Long serves are effective when variation is credible. Overuse trades surprise for predictability. The server must therefore calibrate depth or angle variation not as a secondary option but as a deliberate perturbation within a predominantly short-serve framework.

3. **The third-ball attack: from control to conversion**

The third-ball attack (I_3) is a structured offensive pattern in which a restrictive serve narrows the receiver's options, engineers a predictable return, and enables a pre-planned attacking stroke on the third contact. Its success depends on the precise calibration of serve (spin, placement and tempo), rapidly interpretating the return, and using effective footwork to achieve an optimal attacking position.

This serving pattern is not a tactic but a central organising structure of competitive table tennis. A full treatment is presented in Sect. 4.6.

4. **Return strategies and the adaptive cycle**

Return quality depends on accurate recognition of spin axis and magnitude. Two broad strategies can be distinguished:

- **Neutralisation**: Centralise the contact path to minimise lateral impulse, reducing the effect of incoming sidespin.
- **Spin utilisation**: Redirect the ball along the incoming rotational vector with minimal added force (Sect. 3.4.2), exploiting the server's spin to create angle.

Misinterpretation of compound spin commonly produces over-correction, pop-ups or lateral drift. The receiver's error, however, is not just technical; it is *informational.* Every serve, whether returned successfully or not, generates data about the opponent's perceptual and anticipatory capacities.

Two governing principles follow:

- **Variation**: Diversity prevents predictability and creates opportunities to seize the server's initiative. This principle suggests '*vary every serve*'.
- **Exploitation**: Effective combinations of spin, placement and disguise should be *reused* as repeatable tactical weapons. This principle is also known as '*doing what troubles the opponent*'.

The serve thus represents the highest concentration of tactical control. It may impose dominance through persistent execution of a highly refined pattern or destabilise through calibrated diversity. Here, the distinction between *aggressiveness* and *initiative* becomes critical. Aggressiveness concerns intensity. Initiative concerns control of the evolving interaction. A failed service variation does not imply loss of initiative; it generates information. Initiative is preserved through recalibration. It is forfeited when unsuccessful patterns are repeated without adjustment.

Understanding these tactical patterns requires a recap at the spin architecture that makes them possible, the subject of the next section.

4.4.3 Spin Architecture in Serving

Topspin and backspin define the primary vertical structure of most serves. Sidespin (θ) introduces a third, lateral control axis. When the bat brushes horizontally across the ball's equator, rotation is generated about a near-vertical axis. After table contact, lateral deviation becomes more pronounced during the second flight F_2 (Fig. 2.16) imposes pressure on the receiver.

Pure sidespin is rare in match play. Most advanced serves use compound spin, combining sidespin with either topspin or backspin. The objective is not curvature alone but ambiguity of spin axis and bounce behaviour (Table 4.2).

4.5 Tactical Foundations

Tactics are structured decisions made within a defined playing situation. They determine *when* and *why* a serve or stroke is selected. Technique determines *how* it is executed. The two are inseparable but functionally distinct.

Table tennis is often compared to chess. The comparison is instructive but incomplete. Both reward foresight, deception and initiative. However, table tennis compresses decision-making into tens of milliseconds. Anticipation is therefore not advantageous; it is essential. Points are constructed rather than answered. Players

Table 4.2 Compound spin states-notation and tactical effect

Compound spin type	Symbol	Tactical effect
Topspin-sidespin	$\sigma \mathrm{L}\theta$ or $\sigma \mathrm{R}\theta$	Curved and dipping; rapid lateral acceleration after the bounce
Backspin-sidespin	$\partial \mathrm{L}\theta$ or $\partial \mathrm{R}\theta$	Low, floating trajectory; ideal for short, deceptive serves
Dominant sidespin with weak vertical spin	$\theta\sigma$ or $\theta\partial$	High perceptual ambiguity; vertical bounce difficult to anticipate

implicitly forecast the evolving energy state, spin axis and spatial displacement a few contacts ahead. Tactical competence is therefore predictive, not reactive.

4.5.1 From Stroke Quality to Tactical State

As established in Sect. 2.15, every stroke carries a mechanical signature defined by its stroke quality, $\varphi = (J, S)$, where $J = (J_h, J_b)$ and $S = \omega \cdot R/v$. This quantity describes what is executed at the point of contact. However, stroke quality alone does not produce tactical effect. It becomes tactically meaningful only when translated into a tactical state that the opponent must confront.

The tactical state (T) is defined as:

$$T = (E, G) \tag{4.1}$$

where:

- E = total kinetic energy of the ball ($E = E_v + E_\omega$), arising directly from the stroke quality (J, S).
- G = geometric pressure imposed through placement and angular constraint, $G = g(\alpha, z)$.

The causal chain established in Sect. 2.15 therefore extends naturally:

$$(J, S) \rightarrow E(v, w) \rightarrow T(E, G) \rightarrow \textit{technical pressure}\ \varphi_{opp} \tag{4.2}$$

That is: the player executes a stroke with a certain quality (J, S). This determines the ball's energetic state $E(v, \omega)$. When this energy E is combined with geometric placement G, the ball presents a tactical state T to the opponent. Technical pressure φ_{opp} is then the difficulty that the opponent experiences when attempting to return that ball, which depends on both T and the opponent's adaptive capacity Eq. (2.38).

Technical pressure is therefore not a property of the ball alone. It is the relational consequence of a tactical state imposed on a specific opponent.

4.5.2 Energy and Geometry

At a structural level, tactical decisions operate within two coupled domains: energy (E) and geometry (G). Energy determines intensity. Geometry determines structural difficulty.

The energy domain

As defined in Sect. 2.14.1, the ball's total kinetic energy is

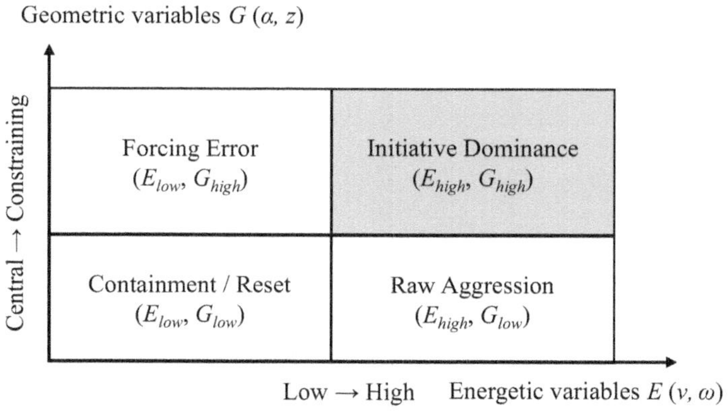

Fig. 4.3 Tactical decision space defined by energy and geometry. The horizontal axis denotes the total kinetic energy of the ball (E) whilst the vertical axis denotes geometric pressure (G) imposed through placement and trajectory angle. Tactical outcomes are shaped by controlled movement within this space rather than by isolated stroke selection

$$E = E_v + E_\omega$$

At each contact, the player selects an energy action. Energy may be injected, redistributed, redirected or depleted:

- **Energy injection**: Topspin, drive and smash increase translational and/or rotational energy, compressing temporal margins.
- **Energy redistribution**: Chop and heavy brush contacts shift energy from translational to rotational modes.
- **Energy redirection**: Block and controlled counter reuse incoming impulse with minimal additional input.
- **Energy depletion**: Push, soft block and *low*-F_h contacts and friction-modifying rubbers (long pips, anti-spin) reduce total system energy.

The resulting spin ratio $S = \omega \cdot R/v$ governs how that energy is distributed between translational and rotational modes.

The geometric domain

The geometric domain is defined by placement (z), trajectory angle (α) and the resulting biomechanical reach constraints. Geometry determines where the opponent must respond and how mechanically efficient that response can be. It therefore imposes structural pressure:

$$G = g(\alpha, z) \tag{4.3}$$

where g is a function that maps spatial parameters to a scalar measure of geometric pressure.

Placement converts energy into spatial demand. By controlling landing location, a player determines the opponent's required displacement, stroke selection and biomechanical stability. Geometry thus transforms abstract energy input into measurable physical demands on the opponent.

Three geometric pressures are particularly significant:

- **Wide placements** increase lateral displacement, extend recovery distance and reduce preparation time.
- **Crossover ('elbow') targeting** forces rapid stroke selection at the transition between forehand and backhand kinetic chains, increasing decision latency and reducing spin production capacity.
- **Short placements** (e.g. small trapezoids) compress forward reach and restrict full-body rotation, forcing operation from mechanically unstable positions.

Empirical zone-based analyses demonstrate that stroke effectiveness varies systematically with table distance and landing position (Munivrana et al. 2015). Identical energy inputs therefore impose different demands depending on spatial application. High energy directed into open space produces time compression. Low energy directed into constrained geometry produces hesitation. Tactical judgement lies in selecting the appropriate energy input and placing it within geometry that disadvantages the opponent (Fig. 4.3).

4.5.3 The Tactical State

Tactical play consists of deliberate movement across the coupled domains of energy and geometry. Each stroke transforms the tactical state T, advancing the rally from exchange n to exchange $n + 1$.

$$T_n = (E_n, G_n) \rightarrow T_{n+1} = (E_{n+1}, G_{n+1}) \tag{4.4}$$

The transformation may involve (Fig. 4.3):

- $E_{n+1} > E_n$ (energy increase)
- $E_{n+1} < E_n$ (energy reduction)
- $G_{n+1} > G_n$ (geometric amplification)
- $G_{n+1} < G_n$ (geometric relaxation).

Effective sequences manipulate both domains simultaneously, adding, redistributing, redirecting, or depleting energy whilst altering spatial demand, to claim or retain initiative.

Consider a structured example:

- **Deep serve to the crossover** → increases geometric pressure: $(E_{mid}, G_{low}) \rightarrow (E_{mid}, G_{high})$.
- **Short push return** → reduces energy: $(E_{mid}, G_{high}) \rightarrow (E_{low}, G_{mid})$.

- **Third-ball attack** → injects energy whilst maintaining geometric pressure: $(E_{low}, G_{mid}) \rightarrow (E_{high}, G_{high})$.

These exchanges are abstracted as:
$(E_{mid}, G_{high}) \rightarrow (E_{low}, G_{mid}) \rightarrow (E_{high}, G_{high})$.

Dominance arises when high energy is delivered into space that is geometrically difficult for the opponent. Although combining high energy with high geometric difficulty (E_{high}, G_{high}) represents the strongest possible outcome (Fig. 4.3), it is difficult to sustain across consecutive strokes. Any loss of energy, placement precision or recovery can force a defensive transition, creating an opening for the opponent to regain initiative.

4.6 Mind in Action

Where Sect. 4.5 formalised the behaviour of the ball, its energy and geometry, this section addresses the internal processes governing the player's response. Tactical models are necessary but insufficient. The player must regulate narrative drift, distinguish execution failure from model failure, maintain directional stability under perturbation, and compress supervisory logic into stable operational states. The focus therefore shifts from the physics of the ball to the player's decision-making.

4.6.1 The Scoreline

Table tennis is a game of razor-thin margins. The scoreline is a compressed transcript of decision pressure, mechanical stability and tactical persistence under extreme time constraint.

A sequence such as

$$11:9, 13:15, 5:11, 11:4, 7:11, 8:11, 11:6$$

signals high volatility. Initiative repeatedly changes hands, often within a persistent technical pattern.

Such oscillations reflect cycles of disruption and recalibration. A player may decode a compound spin axis, recognise a placement bias or stabilise a return pattern, only for that provisional solution to be dismantled within one or two exchanges. Because points are lost quickly, prolonged experimentation is rarely possible. Adaptation must therefore be selective and internally coherent.

Large swings, for example, from 5:11 to 11:4, do not necessarily signal sudden superiority. They often reflect restoration of internal stability. A contained adjustment, such as varying service depth or reasserting pressure on the crossover point, may be sufficient. This phenomenon is universal, observable from amateur clubs to the Olympic stage, and demands a structural explanation rather than emotional interpretation.

4.6.2 *Metaphor and Momentum*

At the highest level, such as in a World Table Tennis men's singles final, rallies unfold near the limits of human reactive capacity. To an external observer, successive strokes appear homogeneous: fast, powerful and interchangeable. Errors, when they occur, resemble stochastic events, for example, a missed return after an extended exchange, or a sudden breakdown within an otherwise balanced rally. From this perspective, match outcomes may appear probabilistic.

The apparent uniformity of rallies invites comparison with a Nash equilibrium in game theory, in which neither player can unilaterally improve outcome given the opponent's strategy (Simon 1955). In professional sport, similar equilibrium-like behaviour has been discussed through mixed-strategy models, where players vary placement and shot selection to minimise predictability and avoid exploitable patterns. Competitive exchanges may therefore appear to operate within a quasi-equilibrium, characterised by mutual adaptation, strategic variation and the temporary absence of clear vulnerabilities (Gauriot et al. 2018 and Ye et al. 2023).

However, experienced play is not sustained at equilibrium but is continually perturbed by execution variability, mechanical constraint and deliberate entropy modulation. Breakthrough arises from its destabilisation: the progressive accumulation of geometric and energetic pressure, the induction of model mismatch, and the eventual failure of the opponent's control under pressure.

The apparent similarity of strokes reflects not the absence of variation, but its *sub-perceptual encoding*: continuous micro-adjustments in placement (z), spin composition (S) and impulse magnitude (J), executed within a severely compressed time window. Outcome differentials arise from the cumulative effect of these adjustments, not from probability (Gauriot et al. 2018).

Within this compressed regime, tactical decisions are shaped not only by biomechanics but also by the *internal narratives* through which players interpret unfolding play. As argued by Lakoff and Johnson (1980), metaphors structure perception and guide action. In table tennis, implicit metaphors may introduce systematic distortions into decision-making:

- **The 'reset' metaphor (momentum discontinuity fallacy)**

This framing treats each rally as an isolated event, encouraging the player to mentally 'start afresh' after each point. However, at elite level, momentum neither purely psychological nor episodic; it is mechanical and informational, encoded in the opponent's calibrated timing, confidence-adjusted risk thresholds, and recent success within specific placement–spin patterns. Disregarding this continuity delays adaptation: the player responds to a prior state rather than the current one, producing clusters of errors that are subsequently misattributed to chance.

- **The 'clutch winner' metaphor (terminal-shot fallacy)**

This framing prioritises the production of a single decisive stroke or serve, particularly under high-pressure conditions such as deuce. At elite level, however, the probability of a clean winner against a balanced opponent is low. Advantage is instead generated through *sequential constraint*: progressively forcing the opponent to operate under deteriorating geometric and energetic conditions. The search for an immediate winner promotes over-acceleration (e.g. excessive J_h) and premature commitment, mechanical deviations that manifest as unforced errors.

An unsuccessful stroke should therefore be interpreted as *state information*, an indication that the current configuration of T (E, G) is suboptimal rather than as evidence that the underlying tactical intention is invalid.

Professional performance therefore requires resistance to *narrative drift*: the tendency for emotional interpretation to override mechanical calibration. Effective play depends on preserving the integrity of key control variables, which are hitting (F_h), brushing (F_b) and placement (z), despite scoreline fluctuations or recent outcomes. This is consistent with the observed emotional neutrality of elite competitors. As world champion **Wang Manyu** notes: "Even when I win, I don't dare smile; I don't know how badly I might lose in the next match".

Such restraint is not suppression. It is *mechanical discipline*: the maintenance of internal conditions necessary for repeatable execution within a dynamically evolving system. What appears, from the outside, as probabilistic volatility is, at the level of performance, a tightly regulated process of continuous adjustment within changing mechanical and tactical conditions.

4.6.3 Self-Stability

A frequently misinterpreted feature of elite performance is *self-stability*: the capacity to maintain internal game stability without succumbing to the *reactionary loop* triggered by an opponent's variations.

Although *watch-and-react* strategies appear theoretically appealing, they are practically unviable at high rally tempos. Decision windows are too narrow. Points are conceded before alternative hypotheses can be tested.

Skilled players therefore operate at a calibrated equilibrium between *bounded reflection* and *tactical insistence*:

- **Bounded reflection** is targeted diagnosis, not wholesale reinvention. The question is binary: Has execution degraded, or has the tactical model ceased to unsettle the opponent?
- **Tactical insistence** applies a selected framework for sufficient duration to propagate pressure, provoke predictable responses and expose structural weaknesses.

In attritional rallies, victory often belongs not to the most inventive player but to the one whose internal model remains stable under stress. Tactical collapse is seldom

Table 4.3 Executional error vs model failure

Failure type	Diagnostic question	Remedial action
Executional	Was the error due to footwork, contact angle or timing?	Stabilise mechanics; retain the current model
Structural	Does the opponent possess a hard counter to the model's logic?	Transition deliberately to a secondary meta-model

a failure of creativity. More often, it results from premature abandonment of a viable model.

1. **Execution failure versus model failure**

Tactical insistence must not become dogmatism. Competitive intelligence lies in distinguishing executional error from structural inadequacy. This diagnostic distinction is formalised in Table 4.3. It requires a clear separation between *validity* (internal logic of the model) and *fidelity* (quality of its execution). Confusing the two leads either to stubborn persistence in a failing model or to chaotic switching that destroys coherence.

2. **Foundations of self-stability**

Self-stability is constructed on:

- A rehearsed repertoire of tactical patterns
- Sufficient physical and technical robustness to execute under fatigue
- Match preparation that supplies multiple initial models rather than a single default plan.

Without this foundation, insistence becomes guesswork and reflection collapses into indecision.

4.6.4 Meta-Models and Tactical Control

The following sections distinguish three levels of tactical control. The meta-model determines whether the player should maintain or change the tactical framework. Initiative describes whether the rally is moving towards the player's preferred conditions. Operational states provide the real-time execution modes through which play is actually carried out. Professional players do not select tactics point by point. They operate within tactical frameworks and regulate those frameworks through a higher-order control structure referred to here as the meta-model (Kolar et al. 2025). Match play therefore becomes a contest of layered control logic, in which higher-order decisions shape sequences of exchanges rather than isolated strokes or points.

1. The three-tier control architecture

This layered structure is summarised in Table 21.
At the *execution layer*, the player applies a primary tactical framework, X. Each return is a function of the incoming ball state:

$$Return = f\ (Incoming\ state).$$

At the coordination layer, bounded reflection continuously monitors whether the opponent is developing an effective counter, specifically whether a hard counter Y is beginning to neutralise framework X (Fig. 4.4).

At the *meta-level*, the meta-model M operates not directly on the physical ball state but on the *performance indicators* generated by the coordination layer:

$$M : \{\text{performance data}\} \rightarrow \{\text{maintain X, abandon X, shift to Z}\}$$

The operational flow is illustrated in Fig. 4.4. This architecture aligns with *bounded rationality* (Simon 1955). In high-speed environments, players cannot optimise exhaustively in real time. Instead, they rely on predefined thresholds and heuristics, satisficing rather than optimising. Framework is maintained so long as performance indicators remain within acceptable bounds; a shift to framework Z is triggered only when those bounds are breached.

2. How the meta-model governs each layer

The meta-model separates kinetic execution from tactical deliberation. At ball speeds approaching 30 m/s, deliberative evaluation between strokes is not feasible. The player cannot consciously analyse every incoming state, compare all tactical options and then select an optimal response. Instead, the meta-model uses predefined trigger rules to govern when a tactical framework should be maintained, modified or replaced

For example, if three consecutive forehand drives are anticipated and neutralised by the opponent, the player may initiate a transition to framework Z, modifying tempo, placement or spin ratio. The shift is rule-based, not emotional. The player changes framework because the current model has begun to fail, not simply because a point has been lost.

Table 4.4 The three-tier tactical control architecture

Layer	Component	Function	Control analogue
Meta-level	Meta-model (M)	Determines tactical framework maintenance or shift	Supervisory control
Coordination	Bounded reflection	Monitors counter-tactical intensity	Error signal evaluation
Execution	Primary framework (X)	Produces dynamic stroke output	System dynamics (e.g. $\theta\sigma$)

3. Countering narrative drift

Players naturally construct causal narratives around error and losing points (Sect. 4.6.2). Without a structured monitoring mechanism, these narratives introduce emotional inference into tactical decision-making. The meta-model reduces this risk by replacing subjective interpretation with tracked variables, such as:

- Variance in landing distributions
- Reduction in ε-strength (Sect. 6.5)
- Repeated exposure to a hard counter.

Tactical transition becomes a *probabilistic adjustment* governed by measured performance degradation, not a fear-based or impulsive reaction.

4. Scheduled transition and system stability

By scheduling framework transitions in advance, rather than reacting impulsively, the meta-model preserves system stability as the external game state evolves. The diagnostic structure is shown in Fig. 4.4 and consists of three linked elements:

- **Primary framework** (***X***): the baseline tactical model applied with insistency.
- **Bounded reflection**: continuous monitoring for the emergence of a hard counter Y.
- **Scheduled transition**: a deliberate shift to an alternative framework Z, designed to exploit the opponent's commitment to Y.

This layered control structure is summarised in Table 4.4.

Operational flow is illustrated in Fig. 4.4:

Failure therefore has two distinct forms. In *fidelity loss*, the player can no longer execute framework X reliably. In validity loss, the opponent's counter Y has rendered framework X ineffective, even when execution quality remains high. The meta-model

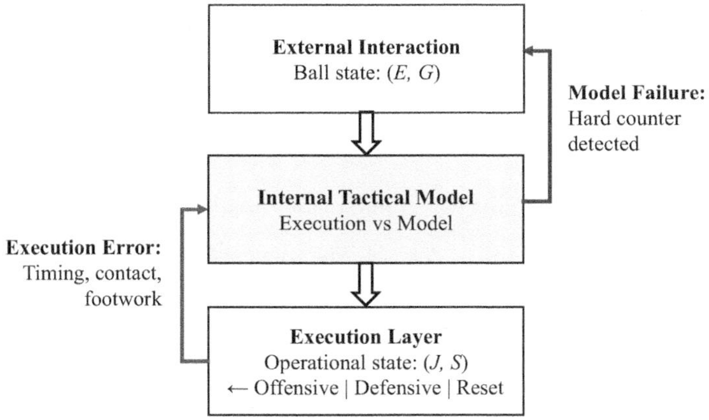

Fig. 4.4 Layered tactical control architecture and diagnostic failure modes. External ball states, expressed through energy and geometry, are mediated by an internal tactical model and realised through execution dynamics. Failure arises either from degraded execution, referred to as fidelity loss, or from invalidation of the current model by a hard counter, referred to as validity loss

distinguishes between these failure types and selects the appropriate response: technical stabilisation when execution has degraded, or tactical transition when the model itself has been invalidated.

4.6.5 Initiative as Directional Stability

1. **From possession to direction**

Initiative is often described informally as being proactively on the attack. This conflates two mechanically distinct ideas. *Aggressiveness* concerns the amount of energy injected into a stroke: how hard the ball is struck and how much spin is applied. *Initiative* is different. It is directional control of the rally's path through tactical space: the capacity to steer successive exchanges towards favourable states, whilst resisting the opponent's attempts to redirect that path.

A player can be highly aggressive and still lack initiative if each high-energy stroke is countered, redirected and absorbed by the opponent. Conversely, a defender can hold initiative through geometric control alone, narrowing angles and reducing tempo, provided those choices consistently pull the rally towards positions from which high-energy is utilised. What matters is not the force applied but who governs the direction of travel.

$$T_1(E_1, G_1) \rightarrow T_2(E_2, G_2) \rightarrow T_3(E_3, G_3) \rightarrow \ldots$$

2. **Perturbation and steering**

Each rally traces a path through tactical space. Initiative exists when a player can sustain movement towards a preferred region despite the disturbances introduced by the opponent's strokes. For an attacker, that region may involve high energy and constrained geometry; for a defender, it may involve lower energy, wider recovery margins and geometry that neutralises attack.

The opponent's stroke acts as a disturbance (ΔT_{opp}) displacing the rally away from the player's intended path.

Initiative therefore requires two simultaneous conditions:

- **Magnitude**: $\|\Delta T_{self}\| > \|\Delta T_{opp}\|$, the player's adjustment overcomes the opponent's disturbance.
- **Direction**: ΔT_{self} moves the state towards the player's preferred region (e.g. E_{high}, G_{low}) rather than being pulled towards the opponent's preferred region (e.g. E_{low}, G_{mid}).

Loss of initiative has two signatures. First, the rally's direction reverses and begins moving away from the player's preferred region. Second, the rally oscillates without convergence, with neither player able to steer consistently. The first is a positional loss; the second is a contest of competing perturbations neither player is yet winning.

3. **Connection to the meta-model**

This is what bounded reflection monitors at the coordination layer (Table 4.4). The meta-model does not wait for initiative to be fully lost before acting. It detects early signs of directional instability and schedules a corrective framework transition before the rally path diverges too far to recover. Incipient loss of initiative is the error signal that triggers supervisory intervention.

Let preferred tactical region be $T_R \subset \{(E, G)\}$. Initiative exists when:

$$T_n \to T_{n+1} \to \cdots \to T_R$$

moves consistently towards T_R despite perturbations.

The energy–geometry sequences analysed in Sect. 4.5.3 illustrate this principle. Transitions such as medium energy with high geometric pressure (E_{mid}, G_{high}), followed by lower energy with moderate geometry (E_{low}, G_{mid}), and then restored high energy with high geometric pressure (E_{high}, G_{high}), show controlled steering through tactical space.

The present section provides the directional logic underlying those sequences: a initiative is sustained movement towards a preferred tactical region against perturbation from the opponent.

4.6.6 Operational States

The meta-model (Sect. 4.6.4) governs when to maintain or shift a tactical framework. The execution layer governs how that framework is delivered, rally by rally. Operational states form the bridge between the two: a small number of stable execution regimes that translate the meta-model's decisions into physical action without requiring conscious deliberation during play. At this layer, decision windows are too narrow and cognitive bandwidth too limited for exhaustive optimisation (Tversky and Kahneman 1974).

1. **Dominant operational regimes**

Three operational regimes are distinguished:

- **Offensive (O_{off})**: Energy is injected and placement variation expanded. Through increased impulse and spin modulation, pressure is actively constructed to constrain the opponent's response space.
- **Defensive (O_{def})**: Placement angles are widened, rally tempo is deliberately varied, and incoming energy absorbed or dissipated. The objective is to force errors.
- **Neutral/Reset (O_{neu})**: Tactical complexity is deliberately reduced. Spin is simplified, placement centralised, and execution recalibrated to prevent error cascades and restore mechanical stability.

These states, or attractor states (Araujo et al. 2025), follow a stability–control–attack sequence,

$$O_{\mathrm{def}} \rightarrow O_{\mathrm{neu}} \rightarrow O_{\mathrm{off}}$$

although reverse or cyclic transitions (e.g., $O_{\mathrm{off}} \rightarrow O_{\mathrm{neu}} \rightarrow O_{\mathrm{off}}$) also occur as control is temporarily relinquished and re-established. The meta-model determines *when* transition is required; the operational state determines *how* each interval is executed.

2. **Why execution can fail independently of tactics**

A tactically sound framework can still collapse at the execution layer. The cause is internal noise: inherent variability in neural firing and muscular activation that, if unregulated, disrupts timing and control (Faisal et al. 2008).

When attention fragments, shifting from executing the shot to judging its outcome or reacting to an error, muscle patterns become less precise and contact timing deteriorates. An internally divided state cannot deliver a unified impulse.

This distinction is diagnostically important. Fig. 4.4 distinguishes two failure modes: *validity loss*, where the tactical framework has been countered and the meta-model must intervene; and *fidelity loss*, where the framework remains sound, but execution has broken down. The second failure originates at the execution layer, and no tactical adjustment from above can correct it. What the player needs is not a new framework, but the internal stability required to deliver the existing one.

3. **Mind as infrastructure**

Drawing on Vickers (2007), the mind is treated here not as a separate emotional variable, but as infrastructure: the underlying architecture that sustains physical performance. Each rally is governed by three interdependent domains:

- Ball ballistics, governed by impulse, spin and energy transfer
- Spatial geometry, governed by placement, angles and biomechanical reach limits
- Internal stability, preservation of conditions required for repeatable execution.

The relationship between tactical opportunity and tactical maintenance is summarised in Table 4.5. This comparison clarifies the distinction between tactical space and tactical control. Tactical space identifies the external coordinates of advantage: energy and geometry, $T\ (E, G)$. Tactical control explains how that advantage is maintained through internal regulation: the stability of the model and the fidelity of execution. The mind, in this formulation, is not an addition to mechanics but the infrastructure through which mechanical intention becomes repeatable action.

Table 4.5 Tactical space versus tactical control

Tactical space (Fig. 4.3)	Tactical control (Fig. 4.4)
Energy × Geometry	Model × Execution
Where advantage lies	How advantage is maintained
Stroke-invariant	Mind as infrastructure
Static map	Dynamic regulation

4.7 The Third-Ball Attack

The *Game Event Sequence* presented in Sect. 2.9.1 allows every rally to be expressed as a chain of impacts (I_n), flights (F_1, F_2) and table contacts (TC). Whilst rallies may extend indefinitely, the first three exchanges ($I_1 \rightarrow I_2 \rightarrow I_3$) disproportionately determine point outcomes, both tactically and statistically. The third-ball attack is built upon this fundamental architecture.

4.7.1 Serve, Return and Third-Ball

Later rally stages (I_{5+}) increasingly resemble probabilistic noise, dominated by improvisation, accumulated error and fatigue. In contrast, the first three impacts form a *deterministic convergence* shaped by mechanical asymmetry:

- **I_1 – Serve (total control)**: Executed from a static state with complete freedom over placement, spin and disguise.
- **I_2 – Receive (forced reaction)**: The post-bounce conditions of F_2, dictated by I_1, sharply limit the receiver's available options.
- **I_3 – Attack (informed aggression)**: The server, now receiving information, attacks a return shaped by engineered I_2.

This asymmetry makes the sequence $I_1 \rightarrow I_2 \rightarrow I_3$ designable. Rather than reacting, players *engineer* the opening exchanges to create advantage *before* the rally becomes fully interactive.

4.7.2 Server Options

A powerful serve (I_1) is the opening move in a deliberate sequence designed to shape the rally state at I_2. The primary lever is placement (Sect. 4.4.2) and the spin ratio

$$S = \omega \cdot R/v,$$

after the first table contact (TC_1).

- **High-S serves (heavy spin, low F_h)**: The server engineered a high-spin, low-F_h ball after TC_1 that is difficult to structure, often inducing hesitant or poorly controlled returns. *Target state*: a high, attackable ball.
- **Low-S serves (high speed, low spin)**: The main limiting factor is the shortened reaction window (t_w), rather than spin handling. The receiver prioritises compact, stabilising contact, producing a low-spin, geometrically predictable return. *Target state*: a neutral but geometrically predictable ball.
- **Deceptive serves (disguised spin axis)**: The aim is to induce a *category error* at I_2 (e.g. misreading backspin as topspin), yielding a popped-up or netted return. *Target state*: a terminally weak ball.

Thus, I_1 is selected not for its direct point-winning probability but for the *predictable class* of I_2 responses it provokes.

4.7.3 Third-Ball Window

As the ball returns to the server ($I_2 \rightarrow F_1 \rightarrow TC \rightarrow F_2 \rightarrow I_3$), the server's prior design crystallises into the third-ball window. Two mechanisms create the opportunity:

- **Geometric restriction**: A deep, spin-laden F_2 restricts the receiver's bat-angle and shortens reaction interval (t_w), increasing temporal and spatial predictability.
- **Energetic favourability**: Weak I_2 returns (too long, too high or poorly timed) provide an ideal energy state for an aggressive stroke.

Experienced players do not *decide* what to do for I_3. They *recognise* which pre-planned branch of their attack matrix has been activated. This creates a predictable oscillation in decision freedom:

- **I_1**: The server acts maximal freedom.
- **I_2**: The receiver acts under severe pressure, with minimal freedom.
- **I_3**: The server acts again with informed freedom, using critical information about the receiver's weakness, timing and pattern.

This asymmetry explains the disproportionate influence of the first three exchanges on rally outcomes.

4.7.4 Countering the Counter

At advanced levels, the receiver is fully aware of being deliberately engineered at I_2. The duel therefore evolves into a layered contest of anticipation rather than a simple execution of serve–receive patterns:

- **Server's primary plan**: My pendulum serve (I_1) will elicit a long push (I_2), enabling a prepared third-ball loop (I_3).

- **Receiver's counterplan**: I will flick early at (I_2), disrupting the server's third-ball timing and collapsing the intended sequence.
- **Server's meta-plan**: I will disguise a no-spin serve using identical motion; the receiver's committed flick will either fail or produce an over extended return that can be countered immediately.

The classic sequence $I_1 \rightarrow I_2 \rightarrow I_3$ thus ceases to be a deterministic pattern and becomes an *informational duel*. Disguise competes with early reading; tactical insistence competes with calculated risk.

Victory does not accrue to the player who reacts fastest but to the one whose internal model operates at a deeper layer, anticipating not only the opponent's action but the opponent's expectation of that action.

4.7.5 Summary: the Third Ball

The third-ball attack transforms play from reactive exchange into a sequence of designed, asymmetric engagements.

By understanding the physics of the convergence ($I_1 \rightarrow I_2 \rightarrow I_3$) and preparing a conditional attack matrix at I_3, a player exploits the rally's inherent structural imbalance. The objective is not simply to attack on the third ball but to ensure that I_3 becomes the logical and mechanical culmination of a sequence deliberately initiated at I_1.

4.8 First Offensive Shot (FOS)

Within the $I_1 \rightarrow I_2 \rightarrow I_3$ convergence, the first offensive shot (FOS) marks the moment when engineered asymmetry is converted into measurable advantage. It is defined here as the first stroke in a rally that does not carry backspin (ə), signalling both the seizure and initial stabilisation of initiative (Fuchs and Lames 2021).

The FOS is therefore not simply the first attack but the tactical pivot at which a rally transitions from conditional preparation to active imposition. All subsequent exchanges inherit their structure from the quality and function of this moment.

4.8.1 The FOS Spectrum

Analysis by Fuchs and Lames (2021), based on 90 matches from the 2016 Rio Olympic Games, identifies four dominant FOS techniques, accounting for 98% of

Table 4.6 FOS techniques in elite play

FOS technique (frequency)	Primary intent	Risk-reward profile
Forehand topspin (37.4%)	Terminal force: win immediately or force an irrecoverable defensive state	High spin and speed; requires time and space. Highest reward, moderate risk
Backhand topspin(29.3%)	Controlled aggression: apply heavy spin from a stable, central (pre-pared) position	High spin with reduced F_h. High reward, low risk (for an attack)
Backhand flip (22.3%)	Tempo seizure: attack a short ball early, denying time	Moderate speed/spin under severe time pressure. Moderate reward, moderate risk
Forehand flip (10.9%)	Surprise and angle creation: exploit wide short balls with unexpected aggression	High angular potential but biomechanically unstable. High reward, very high risk

professional offensive openings (Table 4.6). These techniques are not interchangeable, and each occupies a distinct position along a spectrum of intent, trading off immediacy, spin potential, biomechanical stability and positional consequence.

The observed disparity in winning probability, greater than 50% for topspin-based FOS and less than 50% for flip-based FOS, does not reflect stroke quality but tactical role. The flip functions primarily as a *tempo-stealing intervention* under constraint, whereas the topspin converts previously engineered conditions into decisive force and spin.

In other words, the flip buys time advantage or induces disruption, whereas the topspin cashes that advantage in. Confusing these roles, by demanding terminal outcomes from tempo-seizing strokes or excessive caution from force-conversion strokes, leads to systematic misallocation of risk at the rally's critical pivot point.

4.8.2 FOS and the Serve–Receive Asymmetry

The strategic asymmetry between server and receiver mirrors that introduced at I_1, the origin of the third-ball advantage.

Server-driven FOS (I_3)

The server exploits full control at I_1 to provoke a predictable I_2 return, most commonly a push. The FOS is then realised at I_3 as a premeditated topspin, supported by optimal biomechanical preparation. This pathway represents the highest-percentage conversion of initiative.

The pronounced gender disparity in topspin usage (74.5% versus 91.2%) suggests that professional women's play adheres even more tightly to this design-driven model, emphasising control, preparation and conversion over early disruption.

Receiver-driven FOS (I_2)

The receiver operates within a severely compressed exchange. High-quality serves often deny the time and space required for a full topspin, making the flip, particularly the banana flip, a constrained optimum. It is employed selectively to:

- Pre-empt the server's intended I_3 attack
- Inject uncertainty into a rally designed to be scripted.

The near-even split between flip and topspin among professional male receivers reflects a meta-game in which early aggression is used to punish predictable serve design rather than to impose immediate dominance.

Implication for Training

Empirical evidence shows that over 60% of rallies conclude within two shots of the FOS. The FOS must therefore be understood not as an endpoint but as the first move in a finishing sequence. Training that terminates at a successful FOS is incomplete. Effective drills must explicitly embed the prescribed +1 and +2 shots.

4.8.3 FOS, Self-Stability and Tactical Insistence

The FOS integrates directly with the concepts of *self-stability* and *tactical insistence* introduced in Sect. 4.6.3.

A player's dominant FOS (most commonly a forehand topspin or a no-spin) forms the cornerstone of tactical identity. Insistence consists not in repetition for its own sake but in repeatedly engineering rallies that converge on this opportunity. When an FOS fails, diagnosis must be structured:

- **Executional error**: insist and refine.
- **Pattern failure**: adjust I_1 to re-enable the preferred FOS.
- **Model failure**: transition deliberately to alternative tools (e.g., flip, geometry, tempo change).

This discipline prevents both premature abandonment and dogmatic persistence (Sect. 4.6.4).

4.8.4 Summary: FOS

The first offensive shot is the quantifiable point at which tactical design becomes statistical advantage. Topspin-based FOS represent the primary mechanism for converting engineered conditions into winning probability, whilst flip-based FOS function as essential disruptive tools when initiative is contested.

The competitive objective, therefore, is to systematically increase the frequency and quality of topspin-FOS opportunities, whilst retaining the tactical fluency to contest initiative under pressure. In this sense, the FOS is the operational output of insistent and self-stable game model (Sect. 4.6.4).

The First Offensive Shot also expresses a deeper structural pattern. It marks the *transition from equilibrium to asymmetry:* the moment at which a constrained exchange is opened up through the effective deployment of a weapon, whether topspin, sidespin or pace variation. This transition is the visible onset of initiative.

This 'opening' is created through a designed shot order that incrementally constructs geometric (α, z) and energetic (E_v, E_ω) imbalance, often in forms structurally analogous to the third-ball attack. What appears as a sudden opening is, in fact, the first observable manifestation of a preconditioned sequence.

The FOS is therefore not the cause of victory, but its *first measurable expression*: the point at which a rehearsed tactical architecture begins to assert itself against a resistant opponent. It distinguishes players who just execute strokes from those who systematically construct advantage.

4.9 FORPHEUS Paradigm

FORPHEUS, developed by OMRON (Nakayama et al. 2021 and Asai et al. 2019), is a suspended robotic arm operating predominantly around the centre line of the table. It is neither humanoid nor designed for explosive power. Instead, it offers something far more instructive: a calm, consistent and physics-led model of table tennis, executed without fatigue, emotion or tactical improvisation.

By minimising human limitations such as neurological delay, emotional fluctuation and imperfect anticipation, FORPHEUS exposes the irreducible mechanical core of the game. Its rally model is built not on spectacular winners, but on trajectory prediction, controlled timing and repeatable ball placement (Asai et al. 2019).

4.9.1 Temporal Advantage

FORPHEUS' decisive advantage is temporal. Human players must negotiate the full prediction chain:

$\mathrm{I_n} \rightarrow \mathrm{F_1} \rightarrow \mathrm{TC} \rightarrow \mathrm{F_2} \rightarrow \mathrm{I_{n+1}}$,

where the post-bounce flight F_2, shaped by friction and spin at table contact, is the dominant source of uncertainty and error.

FORPHEUS partially short-circuits this process. Its effective chain is.

$\mathrm{I_n} \rightarrow \mathrm{F_1} \rightarrow \mathrm{TC} \rightarrow$ Interception.

By hitting the ball immediately after the bounce, it compresses the effective F_2 phase and reduces exposure to the region where human prediction is least reliable. The robot therefore operates on a speed-dominant, just-off-the-table ball (τ_{rising}). This approach aligns with its underlying control framework, which continuously predicts incoming and outgoing trajectories whilst accounting for aerodynamic forces and spin-dependent effects (Asai et al. 2019).

4.9.2 *Early Timing*

Early contact is FORPHEUS' defining tactical signature. This confers two primary mechanical advantages.

First, prediction complexity is reduced. Human players implicitly reconstruct both F_1 and the emergent F_2 trajectory, whereas FORPHEUS acts before F_2 fully develops, reducing the interaction to a simpler physical problem. Second, geometry is compressed. Early contact shortens the opponent's response time and stabilises rally tempo into a smooth, medium-paced exchange, precisely the regime in which the robot's centred and controlled motion is most effective.

4.9.3 *Tactical Simplicity and Consistency*

FORPHEUS does not attack with heavy topspin or terminal force. Its game is built on neutralisation: controlled blocks, guided counter-hits and stable placement delivered with exceptional precision.

This simplicity generates pressure in three ways. First, it establishes relentless consistency, creating a quiet rhythmic baseline that exposes impatience and technical instability. Second, it applies pressure through predictability, presenting a clean rally pattern that subtly forces the opponent to introduce asymmetry, often at increased risk. Third, it operates without fatigue or emotional fluctuation. Its tempo neither accelerates impulsively nor hesitates under pressure, making it a stringent test of the human player's patience and internal stability.

4.9.4 *FORPHEUS as a Tactical System*

FORPHEUS embodies highly stabilised closed-loop control in which incoming trajectories are continuously predicted, corrected and returned to a controlled baseline state. Specifically:

- Incoming spin is neutralised through controlled, energy-absorbing impact.
- Geometry is imposed through moderate pace, deep placement and targeting of the opponent's crossover point.
- The system then resets mechanically to an optimal central position, ready to repeat.

This closed-loop stability exploits the physics of the table to its fullest. FORPHEUS does not seek to outplay the opponent through variation. It imposes a structural stability so complete that the opponent confronts an open-loop problem: how to destabilise a system that continuously self-corrects. Risk and complexity are thereby displaced to the human side of the table.

4.9.5 Human Counterstrategy

Defeating FORPHEUS or a human opponent who adopts a similar baseline model demands the deliberate introduction of controlled asymmetry.

- **Disrupt temporal rhythm**: Vary tempo aggressively. Use high-arc, super-long loops to force retreat, followed by sudden drop shots to force advance. FORPHEUS' constant tempo is its vulnerability.
- **Exploit geometric constancy**: Its returns are predictably central. Wide placements should be used not to win outright but to displace the system laterally and then attack the vacated space.
- **Overload spin neutralisation**: Extreme spin magnitudes or rapid spin-axis alternation (e.g. heavy topspin → heavy backspin) can exceed the limits of its soft-block model. The human analogue is *forcing an opponent to abandon a prepared stroke and execute an unplanned one.*
- **Target meta-stability**: A consistency-based system assumes a consistent opponent. Intentional unpredictability in placement, power and spin, whilst maintaining one's own technical stability, is the decisive lever.

4.9.6 FORPHEUS and Self-Stability

FORPHEUS exemplifies the principle of self-stability (Sect. 4.6.3). Its internal model is immutable under variation; rather than adapting reactively to local perturbations, it applies sustained, repeatable pressure that compels the opponent to revise or abandon their own model.

For human players, the lesson is structural rather than imitative. A stable, repeatable baseline must precede all attempts at variation. Only once a stable baseline is established can asymmetry in power, angle or spin modulation be layered deliberately and exploited reliably.

The deeper insight is that FORPHEUS succeeds precisely by being perfectly ordinary. Good table tennis is not defined by spectacular winners but by the ability to force the opponent away from their settled game.

4.9.7 Tactical Lessons for Human Players

FORPHEUS offers several transferable principles:

- Take the ball early (τ_{rising}) to soften spin effects and simplify decision-making.
- Stabilise first through moderate pace, controlled strokes.
- Impose geometry G (α, z) deliberately rather than improvising under pressure.
- Reduce decision complexity by limiting available stroke choices.
- Trust physics: predictable contact leads to controlled delivery of stroke quality (J, S).

Stripped of athleticism and emotion, FORPHEUS exposes the essence of table tennis. Effective play depends on clarity, composure and consistency, not spectacle.

4.10 Entropy in Tactical Play

Tactical advantage in table tennis is generated through the management of energy, geometry, timing and initiative. These elements, developed throughout the preceding sections, now converge into a single unifying meta-principle: *entropy*.

4.10.1 Entropy in a Rally

In physics and information theory, *entropy (E_p) measures the uncertainty or disorder*. A closed system naturally evolves from order towards disorder – entropy increases spontaneously. In a rally, this principle carries a direct tactical implication: play naturally tends toward unpredictability.

A player who accelerates this tendency engineers disorder. The tactical value of a shot lies not only in its speed or placement but in the *informational load* it imposes. The disruptor, who varies spin, pace and placement in ways that resist pattern, systematically degrades the opponent's ability to construct a stable internal model of the rally.

This asymmetry is decisive. A player who imposes high entropy on the opponent whilst maintaining low internal entropy (executing variability from a stable, well-rehearsed technical base) holds a structural advantage.

- **Low entropy (order)**: The rally structure is predictable. One dominant response is available. A sudden down-the-line attack *injects* entropy instantly because the opponent's model has been conditioned by the preceding order.
- **High entropy (chaos)**: The opponent faces multiple plausible outcomes simultaneously under severe time pressure, serve deception, abrupt changes of pace, disguised spin reversal.

1. **Mathematical representation**

The uncertainty experienced by an opponent can be expressed using Shannon entropy:

Table 4.7 Calculating tactical entropy in two serve distributions

Serve direction	Scenario 1, Probability (q_i)	Scenario 2, Probability (q_i)
Wide to forehand	0.10	0.33
The crossover point (elbow)	0.15	0.34
Wide to backhand	0.75	0.33
Entropy (E_p)	**1.054 (bits)**	**1.585 (bits)**

$$E_p = -\sum q_i \times log_2(q_i) \tag{4.5}$$

where q_i denotes the probability of each possible response option. When one reply strongly dominates ($q_1 \approx 1$), entropy is low ($E_p \approx 0$). When several responses are comparably likely ($q_1 \approx q_2 \approx q_3$), entropy is high.

The logarithm is expressed in base 2 following the standard Shannon formulation (Mackay 2005). This representation is also naturally aligned with the compressed decision structure of table tennis. Within a severely compressed time window, the opponent is often forced into immediate binary resolutions: attack or stabilise, commit forward or retreat, prioritise placement or direct pressure. Such rapid, high-stakes branching gives the game an intrinsically binary tactical character, making the base-2 entropy formulation an intuitive measure of competitive uncertainty.

2. **Example: Return of Serve**

Consider a server (Player A) and receiver (Player B). The tactical entropy experienced by Player B depends on the server's placement distribution (Table 4.7).

The balanced server in Scenario 2 imposes nearly maximal uncertainty, forcing the receiver to prepare for all placements simultaneously rather than exploiting a predictable directional bias.

4.10.2 The p0–p1 Duality

Every player, in every rally, is doing one of two things: imposing order or engineering chaos. These two modes, denoted here as p1 (order) and p0 (chaos), represent opposite strategies for shaping the rally's informational environment to gain advantage.

p1: Imposition of order

- **Tactical signature**: Long, spin-heavy serves; structured third-ball attacks; deep power loops and smashes.
- **Mechanical profile**: Economical movements; clean contact; stable, high-output mechanics.
- **Strategic objective**: Create a predictable, high-energy rally where the opponent's available options are narrowed by geometry and handled by rehearsed patterns.

- **Core vulnerability**: Informational rigidity. When incoming spin, pace or geometry falls outside its well-defined response library, decision latency and mechanical breakdown follow.

p0: Architect of chaos

- **Tactical signature**: Short, deceptive serves; varied receives; soft blocks and dead balls; abrupt tempo and spin-axis changes.
- **Mechanical profile**: Compact, adaptable movements with late wrist and forearm articulation; contact designed to contradict visual cues.
- **Strategic objective**: Induce predictive failure, producing hesitation or forced errors exploitable through geometry and placement.
- **Core vulnerability**: Energetic inefficiency. Sustained variation consumes high cognitive and physical resources; under pressure, complexity collapses into error.

Relationship between the poles

- p1 seeks to reduce uncertainty, compressing rallies into repeatable, high-energy exchanges where technical superiority determines the outcome.
- p0 seeks to maximise uncertainty, destabilising the opponent's predictive models whilst maintaining concealed internal stability.

Neither p0 nor p1 is a fixed identity. Players adopt, abandon and switch between them mid-match as scoreline, fatigue, and the opponent's tendencies shift.

4.10.3 Entropy and the Meta-Game

A match is a dynamic struggle to steer the rally towards a preferred entropy state. It is a contest over whether structure consolidates or disorder persists, and over which player controls the direction of that convergence.

- **p1's winning condition**: Suppress, absorb or overpower p0-induced chaos through superior spin-reading, relentless rhythmic insistence and simplifying geometry back to direct energy exchange.
- **p0's winning condition**: Prevent order from stabilising through continuous rhythm denial, disruption of the opponent's preparation, and exploitation of any mechanical or positional rigidity.

Players do not simply maximise entropy. According to the principle of bounded rationality, they avoid exhaustive complexity that would exceed their own cognitive bandwidth (Simon 1955). Excessive disorder reduces control and may destabilise the initiator. Instead, skilled play regulates the timing of entropy, injecting disruption (p0) during certain phases, then reducing entropy (p1) to stabilise the rally and consolidate positional advantage. This controlled alternation aligns with the Stroke Performance

Relevance (SPR) model, where players move between states of high-pressure disruption and stable equilibrium to maximise match-play efficiency (Ye et al. 2023). This duality represents the operational expression of the p0–p1 framework.

4.10.4 Operational Range

Players are not fixed at either p0 or p1. They are intentional navigators, each with a natural centre of gravity, learning to move along the continuum in response to score, opponent and momentum (Fig. 4.5).

For a p1-centred player:

- Consolidate loop quality, spin strength and directional control, making the p1 core sufficiently forceful to crush mediocre chaos.
- Add one or two high-percentage p0 tools (short serves, drop shots, sidespin pushes) to prevent opponents from settling into a counter-rhythm.

For a p0-centred player:

- Refine consistency and penetration of disruptive tools.
- Develop one decisive p1 weapon (e.g. a reliable counter-loop or mid-distance smash) to terminate points once asymmetry is created.

Progress comes through expansion, not replacement: strengthening one's core whilst selectively acquiring counter-qualities from the opposite pole. The continuum functions as both a diagnostic framework for understanding player–style interactions and a developmental map.

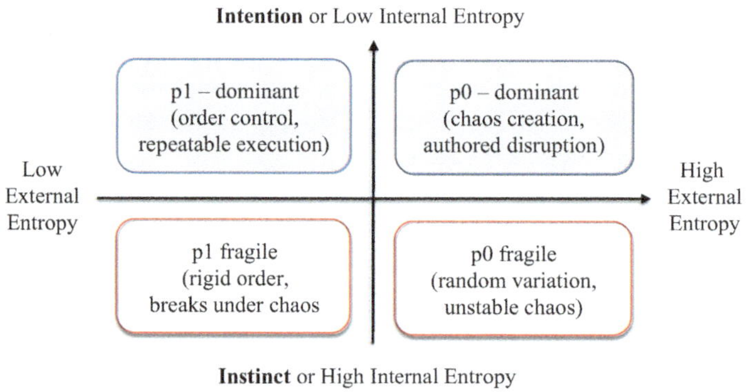

Fig. 4.5 The p0–p1 tactical spectrum expressed as entropy control and operational range. The horizontal axis represents external entropy manipulation: p1 reduces and orders entropy, whereas p0 injects and destabilises it. The vertical axis represents internal entropy capacity, ranging from instinctive fragility to intentional stability under variation

This continuum provides the interpretive foundation for the discrete operational map developed later in Sect. 8.4.5. Whereas Fig. 4.5 describes the player's expandable range along the p0–p1 spectrum, Fig. 8.5 formalises this range into four regulatory states according to whether entropy is governed instinctively or intentionally.

This explains why certain encounters feel structurally unwinnable: one tactical system suppresses the other, shifting the analysis from *"He played better."* to the more precise observation that *"His system invalidated mine."*

4.10.5 Four Layers of Entropy

Entropy in table tennis arises across four distinct levels, as presented in Table 4.8.

These layers form a structured hierarchy in which each level modulates the entropy of the one beneath it. Physical variability provides the substrate of play: even elite athletes must continuously regulate the intrinsic variability of the ball–bat–air system (Faisal et al. 2008). Stroke execution then transforms this variability through the balance between translational and rotational motion, governed by impulse and spin ratio (J, S). Skilled players manage this technical entropy by increasing both speed and spin whilst preserving stability and placement precision (Tian and Xiao 2024). Tactical entropy acts as a higher-order regulator, deliberately amplifying or suppressing variation through timing (τ), placement (z), angle (α) and spin manipulation. This functional variability maintains internal stability whilst denying the opponent an exploitable pattern. Informational entropy represents the runtime consequence of the preceding layers: the uncertainty imposed on the opponent's decision process. Following Shannon's framework (Simon 1955), increasing the entropy of the tactical state forces the opponent to process greater informational complexity within a severely constrained reaction window. Competitive advantage is therefore gained not from entropy alone, but from controlling how these layers interact, stabilise and destabilise one another under pressure.

Table 4.8 Layers of entropy

No	Layer	Description
1	Physical entropy	Intrinsic variability of the ball–bat–air system; dual-force (F_h, F_b), motion (v, ω)
2	Technical entropy	Balance between translational and rotational motion through impulse and spin ratio, i.e. (J, S) where $J = (J_h, J_b)$, $S = \omega \cdot R/v$
3	Tactical entropy	Deliberate variation in placement, spin and rhythm, $T(E, G)$ *where* $E = (E_v, E_\omega)$, $G = (\alpha, z)$ and timing (τ)
4	Information entropy	Uncertainty imposed on the opponent's decision process

Table 4.9 Internal entropy model and external entropy goal

Engine	Internal model	External goal
p1 (the reducer)	High-capacity processor optimised for pace-dominant signals; tolerates fast, spin-rich, high-energy input	Drive the rally's external entropy toward zero, collapsing the opponent's decision space into predictable branches
p0 (the injector)	Wide-bandwidth receiver calibrated for low-impulse, high-variability signals; robust to weak, spinless or rhythmically disrupted input	Maximise the opponent's external entropy through contradiction, disguise and rhythm disruption; flood the opponent's processor with noise

4.10.6 The Entropy Engine

The p0–p1 continuum can be formalised through entropy. Each player operates with an *internal entropy model* (the range of variability they can process reliably) whilst simultaneously seeking to impose an *external entropy state* on the rally. Victory is achieved by forcing the opponent into a state that their internal model cannot efficiently process.

The operational poles of p0 and p1 can be defined as distinct types of *entropy engines*: systems designed either to reduce or to inject uncertainty into the rally (Table 4.9).The player whose internal model remains stable under the imposed external entropy will prevail.

4.10.7 Pimpled Rubber Entropy

The ultimate formal example of a p0 entropy engine is pimpled rubber. It functions as a dedicated, physics-level weapon by systematically decoupling the informational content of a shot from its kinetic execution. It violates the fundamental assumptions on which the inverted-rubber internal model is built by decoupling:

- **Information from action**: Spin is often reversed, rendering the visual cue of the opponent's swing misleading.
- **Energy from outcome**: A powerful swing can produce a dead ball with minimal returnable kinetic energy for the opponent to exploit.
- **Temporal rhythm**: A consistent, high-energy rally cannot be sustained; instead, point-to-point rhythm is repeatedly disrupted.

The result is forced, continuous algorithmic switching by the opponent, driving execution instability even in technically superior players. Here, sophistication loses value; the stability of execution under informational assault becomes the decisive factor.

This mechanism should be distinguished from unstructured disruption: the same decoupling of cue, spin and outcome may also appear incidentally in soft, irregular or p00-style play, but without the control required to convert entropy into sustained tactical advantage. This developmental form is examined in Sect. 8.4.1.

4.10.8 Resolving Style Clashes

If disorder increases spontaneously, why does order ever prevail? Because entropy imposed on an opponent confers advantage, whereas entropy within one's own execution is catastrophic. Experienced players externalise disorder whilst maintaining internal order.

- **Thesis (p1)**: Order through technical mastery.
- **Antithesis (p0)**: Disruption through variation and rule-breaking.
- **Synthesis (adaptive fluency)**: Navigate the entropy spectrum fluidly, deploying order or chaos as demanded, whilst maintaining internal stability.

The synthesis arises through three convergent mechanisms:

- **Overwhelming force**: A p1 system of sufficient magnitude compresses entropy by eliminating time to process it.
- **Expanded internal models**: A p1 system with modular adaptations for unconventional input (long pimples, heavy sidespin, tempo shifts) absorbs disorder without destabilisation.
- **Sustained initiative**: Control of the first offensive shot (FOS) determines the initial entropy state of the rally.

Practical navigation

The golden principle: varying the game continuously and making every rally unsettling, never allowing the opponent to settle into a rhythm, is supported by entropy theory but must be applied asymmetrically depending on playing style:

- **p1 facing chaos**: Simplify the interaction first. Impose deep, spin-dominant geometry; overwhelm external entropy through sustained, predictable force. Order eventually exhausts chaos.
- **p0 facing power**: Disrupt pre-emptively, not reactively. Deny preparation; disrupt serve and receive before the p1 player establishes rhythmic control. Chaos introduced early is exponentially more effective.

In both cases, the decisive factor is not the tactic itself but the timing and consistency of its application. Entropy compounds: small, sustained variations accumulate into structural advantages that become increasingly difficult to reverse (Ye et al. 2023).

4.10.9 Creating Entropy

Table tennis ultimately becomes a contest over the entropy state of the rally. At the highest levels of play, the decisive player is not constrained by either order or chaos but regulates both, rendering the rally either transparent and predictable or deliberately opaque to the opponent.

In this sense, table tennis becomes a contest over the informational structure of the rally. By manipulating spin, placement, timing and pace, players expand the opponent's uncertainty whilst stabilising their own execution. At the highest level, the player is no longer subject to the rally's entropy but becomes *its creator*.

References

Araujo, D. et al. (2025). The ecological dynamics of cognizant action in sport. Psychology of Sport & Exercise 80 (2025) 102935. https://doi.org/10.1016/j.psychsport.2025.102935

Asai K, Nakayama M, Yase S (2019) The ping pong robot to return a ball precisely: trajectory prediction and racket control for spinning balls. Omron Tech 51(1):174–179. Technical Research Paper (Timing/Anticipation). https://www.omron.com/global/en/technology/omrontechnics/vol51/016.html

Djokic Z et al. (2013) Facta Univ: Ser Phys Educ Sport 11(3), 275–284. https://doi.org/10.3390/app1209426

Fuchs K, Lames M (2021) First offensive in Elite Table Tennis. Int J Racket Sport Sci 3(1). https://doi.org/10.30827/Digibug.70278. (First Offensive Shot in Elite Table Tennis)

Faisal A. Aldo, Luc PJ Selen, Daniel M. Wolpert (2008) Noise in the nervous system Nature Reviews Neuroscience 9(4) 292-303. https://doi.org/10.1038/nrn2258

Gauriot R et al. (2018) Nash at Wimbledon: Evidence from half a million serves, https://www.johnwooders.com/papers/NashAtWimbledon.pdf

The International Table Tennis Federation (ITTF) (2026) *Statutes*, https://documents.ittf.sport/sites/default/files/public/2026-02/2026_Statutes_v1_consolidated_clean.pdf

Tian J, Xiao Y (2024) Research on the difference of stroke characteristics and stroke effect between different stroke duration of table tennis players Scientific Reports 14(1). https://doi.org/10.1038/s41598-024-76802-2

Kolar E et al. (2025) Conceptual framework of coaches' decision-making in conventional sports. Front Psychol 5. https://www.frontiersin.org/journals/psychology/articles/10.3389/fpsyg.2024.1498186/full

Lakoff G, Johnson M (1980) Metaphors we live by. University Chicago Press. ISBN: 978-0226468013

MacKay DJC (2005) Information theory, inference, and learning algorithms. Chapter 2 – Stream codes, Equation (6.16), *Cambridge University Press*, Version 7.2, https://www.inference.org.uk/itprnn/book.pdf

Munivrana G et al (2015) Structural analysis of technical-tactical elements in table tennis and their role in different playing zones. J Hum KinEt 47:197–214. https://pmc.ncbi.nlm.nih.gov/articles/PMC4633256/

Nakayama M et al (2021) Ping-pong robot to control motivation of a human player. OMRON TECHNICS 53:008EN. https://www.omron.com/global/en/technology/omrontechnics/vol51/016.html

Simon HA (1955) A behavioral model of rational choice. Quart J Econ 69(1):99–118. https://doi.org/10.2307/1884852

Tversky A, Kahneman D (1974) Judgment under uncertainty: heuristics and biases. Science 185(4157):1124–1131. https://doi.org/10.1126/science.185.4157.1124

Vickers JN (2007) Perception, cognition, and decision training: the quiet eye in action. ISBN: 9780736042567

Xu L et al (2022) Perceptual judgments for table tennis serve recognition: an event-related potentials study. J Appl Sci 12(9). https://doi.org/10.3390/app12094261

Ye R et al (2023) Nash equilibrium and tennis serve performance: a game theory analysis. Int J Perform Anal Sport 23(6):515–526. https://doi.org/10.1080/24748668.2023.2256120

Chapter 5
Dialectical Understanding of Attributes

Table tennis is shaped by a series of persistent oppositions: speed and spin, hitting and brushing, stability and disruption, instinct and intention. These are not binary choices to be resolved but dynamic tensions whose interaction drives tactical depth and adaptive play. Dialectical thinking provides a structured analytical framework for understanding these tensions as mutually shaping forces rather than contradictions to be eliminated.

Across this study, dialectics functions not as abstract philosophy but as a conceptual engine that unifies the tactical systems already developed, from active and reactive returns, self-stability and entropy management to the p0–p1 continuum.

5.1 Contradictions in Play

The analysis of table tennis mechanics consistently reveals a recurring dual structure among key attributes:

- **Speed versus Spin**: Speed applies pressure by compressing reaction window (t_w); spin disrupts by reshaping flight and bounce trajectories.
- **Hitting force versus Brushing force**: Normal impact force (F_h) transfers momentum directly into the ball, producing speed and penetration; tangential brushing force (F_b) generates spin, altering trajectory, bounce behaviour and rally stability.

Just as effective tactics require both a mechanism for initiative and a mechanism for protection, high-level play depends on navigating these opposing forces within a rapidly evolving rally. This navigation is not a compromise between poles but a synthesis that produces a higher operational capability.

X. Ji, *Spin, Speed and Strategy*, Mathematical Engineering,
https://doi.org/10.1007/978-3-032-25518-1_5

Applying Hegelian Dialectics

To formalise this process, Hegelian dialectics (Hegel and Di Giovanni 2010) provides a productive framework. Dialectics describes development as the interaction of a *thesis* and its *antithesis*, whose tension generates a more complex *synthesis*.

Applied to table tennis, this framework explains why certain oppositions organise tactical possibility rather than simply restrict it:

- **Thesis**: an initial tactical tendency (e.g. speed-dominant play).
- **Antithesis**: a counterforce that disrupts or negates it (e.g. heavy backspin).
- **Synthesis**: a higher-order resolution (e.g. the fast, heavy-spin loop).

This dialectical motion recurs throughout the sport. It explains the evolution from instinctive anchoring (p0 or p1) towards intentional control and why style clashes are resolved not through persistence alone but through adaptive synthesis. In practice, these dialectical tensions do not manifest continuously but condense into a limited number of stable control patterns that can be identified and formalised.

The following section uses the fundamental Hit–Brush (F_h–F_b) tension to make this structure explicit.

5.2 The Hit–Brush Dialectic (F_h versus F_b)

Few dualities are as mechanically fundamental as the tension between normal force (hit, F_h) and tangential force (brush, F_b). Every stroke is a particular balance of these two components.

1. **Thesis: Hit as quantity (F_h)**

Hit represents the *quantitative* dimension of stroke production. Techniques such as driving, punching and flat hitting prioritise translational velocity (v), aiming to compress the opponent's response window (t_w). By injecting linear kinetic energy (E_v), Hit produces high-energy trajectories that overwhelm through pace.

This is the mechanical essence of the p1 player's game.

2. **Antithesis: Brush as quality (F_b)**

Brush emphasises qualitative transformation. Through tangential impulse J_b, rotational velocity ω and spin ($\sigma, ə, \theta$) are generated, reshaping flight via drag and Magnus effects and modifying bounce through spin transformation.

Brush introduces variation and deception. It is the primary mechanism of intention and the injection of entropy. This constitutes the core toolset of the p0 disruptor.

3. **Synthesis: unity and conflict**

Hit and Brush are not mutually exclusive. Pure Hit risks predictability; pure Brush risks insufficient penetration. Neither exists in isolation: even flat hits require minimal brush for stability and even heavy loops require sufficient normal force to stabilise trajectory.

Their conflict generates tactical richness, whilst their unity underpins advanced stroke mechanics. The elite topspin loop, combining high F_h and high F_b, is the paradigmatic synthesis. It imposes order by uniting overwhelming force with disruptive spin.

4. **Transformation and context**

Player development often begins anchored at one pole and evolves through a synthesis. With experience, players learn:

- To hit with intention, adding spin (F_b) for safety and shape.
- To brush with conviction, adding speed (F_h) for penetration.

The Hit–Brush balance is always contextual. Opponent style, score pressure, energic and geometric states determine which component must dominate. This adaptive balance is the practical expression of self-stability: knowing when to insist on one's synthesis and when to adjust its weighting.

5.3 Quantifying the Dialectic

Dialectics provides the conceptual architecture; positivism, as articulated by Auguste Comte (Comte 1855), supplies the methodological discipline required to operationalise it. Positivism asserts that valid knowledge must be observable, measurable and verifiable.

Applied to table tennis, this perspective shifts dialectical pairs: Hit versus Brush, Speed versus Spin, and Initiative versus Reaction, from metaphorical contrasts to measurable physical variables. Linear and angular velocities (v and ω), together with normal and tangential forces (F_h and F_b), are treated as empirical quantities rather than qualitative descriptors, as *they naturally are*.

Within this framework, the tension between speed and spin becomes an objective relationship, captured by the spin ratio $S = \omega \cdot R/v$.

This formalisation justifies the introduction of structured binary notation as a compact language for describing play:

- Impact action: Hit (1) versus Brush (0)
- Motion priority: Speed (1) versus Spin (0)
- Energetic state: High-Energy (1) versus Depleted-Energy (0)
- Tactical return: Active (1) versus Reactive (0)
- Strategic intent: Offence (1) versus Defence (0)
- Systemic state: High-Entropy (1) versus Low-Entropy (0)
- Player profile: p1 Order (1) versus p0 Chaos (0).

The use of binary notation provides the mathematical basis for the Table Tennis Notation System (TTNS) developed in later chapters.

Within a positivist framework, synthesis is not assumed to be superior; it must be demonstrated. Empirical analysis of the first offensive shot provides such validation.

The topspin FOS, representing a synthesis of F_h and F_b, exhibits a winning probability exceeding 50% (Sect. 4.8.1), thereby establishing its structural superiority over isolated hitting or passive brushing alone.

5.4 Developmental Synthesis

The dialectical framework sharpens tactical understanding in three domains:

1. **Decision-making under pressure**

Framing choices as tensions rather than absolutes enables recalibration within the rally. Instead of rigid commitment to order (p1) or chaos (p0), players shift emphasis according to trajectory, geometry, energetic state and score. Dialectics converts either/or decisions into continuous adjustment along the p0–p1 continuum.

2. **Player development**

Development is a process of synthesis, not replacement (Fig. 4.5):

- Instinct integrates intention.
- Disruption acquires stability.
- Speed learns to use spin as control.

This provides a structured roadmap for player development: expansion from a stable core rather than abandonment of identity.

3. **Analytic clarity for TTNS**

Dialectics aligns naturally with TTNS discretisation (Chaps. 6–10). Binary codes encode opposing tendencies, whilst transitions and combinations represent synthesis. Contradictions become measurable patterns rather than vague intuition.

5.5 The Philosophy of Mastery

It is unsurprising that table tennis has attracted profound philosophical reflection. The game demands extreme precision, its governing logic is subtle, distributed and often hidden. Outcomes hinge not only on biomechanics but on the regulation of initiative, the propagation of energy within rallies, the management of uncertainty, and the stabilisation of intention under pressure.

However, the value of philosophical traditions in table tennis does not lie primarily in telling players *what to do*. Rather, these traditions cultivate *how to see*: how to interpret struggle, error, rhythm, fatigue and opposition without becoming captive to them. They slow judgement, deepen reflection and integrate physical practice with mental and ethical development.

Table 5.1 Philosophical traditions in table tennis

Philosophical tradition	Reflective contribution to practice
Daoism	Emphasises non-resistance, allowing the opponent's force to dictate movement and returning it with minimal interference. This is the principle behind the reactive return, disruptor's (p0) adaptability and the FORPHEUS model of neutralising rally complexity
Buddhism	Develops non-attachment to outcome and identity. Errors, points and matches are experienced fully but not clung to. This cultivates resilience, presence and the capacity to continue without internal fragmentation. This is the foundation for self-stability and tactical insistence, preventing collapse after errors
Confucianism	Frames improvement as ethical self-cultivation through disciplined repetition and reflection. Mastery is gradual, earned and rooted in respect for process rather than display. This supports consistency and long-term development
Stoicism	Clarifies the boundary between internal control and external uncertainty. By accepting what cannot be governed, e.g., opponent behaviour, luck, momentary fluctuations, the player preserves internal stability and avoids emotional overreach

In this sense, table tennis functions as a life practice. The table becomes a focused arena in which broader human capacities are exercised: patience under provocation, clarity under speed, discipline without rigidity and composure without detachment. Philosophy does not replace technique, nor does it optimise tactics directly. Instead, it provides a language for understanding why certain behaviours endure, why others collapse and how long-term stability is sustained.

The traditions summarised in Table 5.1 should therefore be read reflectively rather than used instrumentally. They illuminate attitudes and, over time, shape how a player trains, competes, recovers and develops.

These philosophies converge on a common principle: responsibility for action remains internal, while outcomes remain contingent on interacting conditions. Philosophy does not make a player immediately more skilful at the table; rather, it makes the player wiser in relation to effort, failure and persistence. Fitness therefore becomes not just physical conditioning but an integration of body, attention and restraint.

The technical systems developed in earlier chapters describe *how* advantage is engineered. Philosophy explains why the capacity to apply those systems repeatedly, calmly and honestly is rare, and why mastery, in table tennis as in life, is ultimately a matter of sustained consistency rather than isolated brilliance.

References

Comte A (1855) A general view of positivism. Translated by J. H. Bridges. eBook is published in 2015. https://doi.org/10.4324/9781315645780. ISBN: 978-1315645780

Hegel GWF, Di Giovanni G (2010) The science of logic. Cambridge University Press. (eBook, accessed via Manchester Metropolitan University Library)

Chapter 6
Discretising Table Tennis: Foundations of TTNS

This chapter introduces the core methodological step: discretising table tennis mechanics. Continuous physical variables are quantised into a small number of *operational attributes* (Hit versus Brush and Speed versus Spin), turning strokes into discrete, countable events. Building on the decomposition of play into symbolic stages ($I_1 \rightarrow F_1 \rightarrow TC \rightarrow F_2 \rightarrow I_2$, Sect. 2.9), this approach enables more precise analysis of strokes, serves and tactics, and marks the transition from mechanical description to formal representation.

The Table Tennis Notation System (TTNS) is developed for three related purposes: to clarify the internal structure of strokes, serves and rallies; to support player and coach understanding under the time pressure that real competition imposes; and to provide a representation suitable for computational analysis, including AI-compatible encoding. These purposes are complementary: a notation precise enough for computation is also precise enough to sharpen human understanding.

6.1 Continuous and Discrete Quantities in Table Tennis

The mechanical framework established in the preceding chapters (Chaps. 2 and 3) describes how contact forces govern ball dynamics: the through-centre hitting force (F_h) produces forward speed (v), whilst the tangential brushing force (F_b) generates spin (ω). Although these quantities are continuous in physical terms, tactical perception in play operates through categorical distinctions.

Velocity, for example, is rarely interpreted as a numerical value such as 18.4 m/s but as *exceptional speed* requiring a specific tactical response. Similarly, spin is not perceived as a rate of 92 rps but as *heavy spin*, prompting an adjustment of bat angle and timing.

To analyse the game systematically, these continuous mechanical quantities are recast as *discrete attributes*. Discretisation creates a vocabulary for recording and

X. Ji, *Spin, Speed and Strategy*, Mathematical Engineering,
https://doi.org/10.1007/978-3-032-25518-1_6

Table 6.1 Fundamental attributes: hit, brush, speed and spin

Physics quantity	Definition	Discrete attribute	Code	Note
Hitting force, F_h	Through-centre impact	Hit	1	Hit–brush forms the *first fundamental pair* (*contact actions*).
Brushing force, F_b	Tangential impact	Brush	0	
Travelling speed, v	Ball's speed	Speed	1	Speed–spin forms the *second fundamental pair* (*resulting motions*).
Rotating speed, ω	Ball's rotation rate	Spin	0	

comparing events quantitatively. It represents the crucial step *from observing the game to analysing it.*

The following sections formalise two fundamental conceptual pairs derived from the Hit–Brush dual-force model, organised into a binary structure and summarised in Table 6.1.

- **Contact actions**: Hit (F_h) and Brush (F_b)
- **Resulting motions**: Speed (v) and Spin (ω)

6.2 Tokenised Shot Actions and DQ4

To bridge abstract forces and perceived technique, the discrete attributes are mapped onto four ordered, tactically meaningful levels, termed Discrete Quantisation (DQ4). This system prioritises immediate interpretability for players and analysts whilst preserving structural consistency, translating the player's kinaesthetic sense into an analytical code.

1. **Tokenised shot actions**

- Hit (F_h): Nudge (1) → Punch (11) → Smash (111) → Strike (1111).
- Brush (F_b): Caress (0) → Graze (00) → Skim (000) → Sweep (0000).

These mnemonic anchors link binary notation to felt execution, intensity and intent.

2. **The DQ4 scale**

The DQ4 system assigns binary strings to these ordered levels, forming the core of the TTNS notation framework (Table 6.2).

For example, a ball with very high translational velocity is categorised as *exceptional fast (1111)*, rather than defined by an exact numerical value. Thresholds for each level must be defined *a priori* and applied consistently. Two complementary approaches are possible:

Table 6.2 DQ4 scale for speed and spin

Attribute	DQ4 code	Characteristic anchor
Hit	1/11/111/1111	Nudge/Punch/Smash/Strike
Speed	1/11/111/1111	Slow/Fast/Very Fast/Exceptionally Fast
Brush	0/00/000/0000	Caress/Graze/Skim/Sweep
Spin	0/00/000/0000	Slow/Strong/Very Strong/Eexceptionally Strong

- **Absolute thresholds**: Fixed numerical cut-offs derived from measurement or training data.
- **Relative thresholds**: Calibrated to the current pairing of players, reflecting the typical speed and spin levels of both.

Absolute thresholds are appropriate where physical extremes are clearly identifiable. Exceptional speed or maximal force of a *strike* stroke, for example, can in principle be measured using high-speed cameras, radar or instrumented systems. In practice, however, such extremes are more often recognised qualitatively than numerically: a *cleanly timed, fully committed strike* or the characteristic sensation of using the bat *like a hammer* to hit the ball. These events occupy the upper tail of the performance distribution and are perceptually unambiguous to experienced players, even without explicit measurement.

Relative thresholds, by contrast, capture the tactical reality of competitive play. What constitutes fast, very fast or exceptionally fast is not absolute but opponent dependent. A smash that overwhelms one defender may be routine for another with superior anticipation or blocking skill. Similarly, a spin level that is *moderate* for a high-level looper may function as *strong* or even *disruptive* against a player whose internal model is calibrated to lower rotational input.

The TTNS therefore permits both interpretations. Absolute thresholds anchor the notation to physical reality and prevent drift, whilst relative thresholds preserve tactical meaning within a given rally interaction.

6.3 Hybrid Actions

All effective strokes are hybrid actions that synthesise Hit (F_h) and Brush (F_b) actions in varying proportions to produce a desired ball state and flight behaviour.

For example:

- Combining *Punch* (11, Hit Level 2) and *Graze* (00, Brush Level 2) yields the token 1100, which corresponds mechanically and tactically to a fast topspin stroke (a loop).
- Reversing the emphasis, *Graze* followed by *Punch* produces 0011, which behaves as a backspin *Slice* (Fig. 6.1).

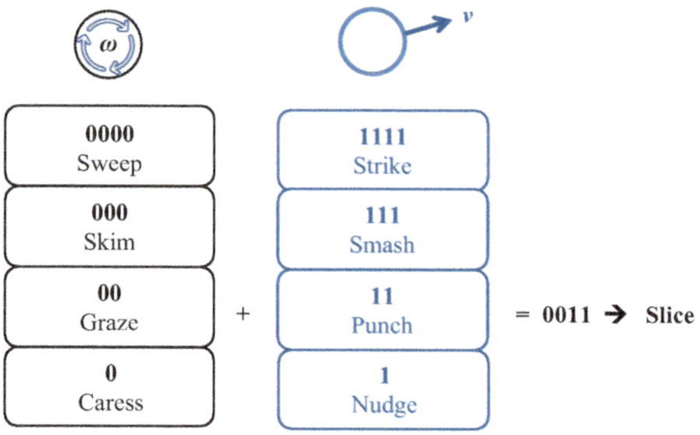

Fig. 6.1 A TTNS symbolic representation of the slice (0011) as a hybrid action combining graze (00) and punch (11)

These hybrid actions validate the binary model, demonstrating how discrete categories recombine to represent the richness of real strokes whilst preserving analytical clarity. TTNS represents stroke techniques using binary strings of one to four digits. The ordering and grouping of these digits encode the relative dominance of Hit and Brush. In many cases, the code is intuitively readable; the full coding rules are explained later in Sect. 8.2.4.

The tokenised *primitives* of the hitting force (nudge, punch, smash, strike) and of the brushing force (caress, graze, skim, sweep) each corresponds to a distinct configuration of stroke quality (J, S). Every discrete action, ether primitive or hybrid, is already embedded with tactical intent. Discretisation thus exposes the game at its mechanical core.

The capacity to blend controlled Hit and Brush components, producing strokes such as topspin (1100) or slice (0011), is central to rally construction and stroke quality, consistent with the principle of mechanical completeness established in Sect. 4.3.

Those tokenised actions (including the hybrids) represent a departure from traditional coaching. Rather than prescribing specific body motions, it specifies physical intent and outcome: the requisite balance between hitting and brushing that defines a given ball state. How such skills are achieved is intentionally left unconstrained. A player's technique may develop through varied mechanics, timing strategies or biomechanical solutions.

The discrete framework thus provides a theoretical layer: a language for describing, analysing, and comparing strokes and techniques. Its role is to formalise what is already present in high-level play, not to prescribe how it should be taught.

Table 6.3 Technical and mental dimensions of the four fundamental attributes

Attribute	Technical characteristic	Cognitive/tactical characteristic
Hit	Flat through-Centre contact, generating forward momentum. Progresses in intensity: Nudge → punch → smash → strike.	Direct and action oriented. Prioritises immediacy and initiative, when dominant, decision-making tends to be rapid, and commitment driven. This operating mode commonly underpins the p1 player's direct J_h exchanges.
Brush	Tangential contact (upwards, downwards or lateral), generating spin and curvature. Progresses in intensity: Caress → graze → skim → sweep.	Intentional and adaptive. Emphasises manipulation, variation and control. This mode supports entropy injection and is central to p0-style disruption.
Speed	Linear forward motion that compresses the opponent's reaction window. Progresses from slow → fast → very fast → exceptionally fast.	Produces urgency and temporal pressure, often forcing abbreviated or defensive responses.
Spin	Rotational velocity that injects high entropy and unpredictability. Progresses from light → strong → very strong → exceptionally strong.	Disruptive and deceptive. Alters predictability and governs rally tempo.

6.4 The Mind-Body Link

Table 6.3 maps the four fundamental physical attributes of strokes to their corresponding cognitive and tactical tendencies. This separation clarifies how mechanical properties translate into tactical intention, whilst also revealing characteristic vulnerabilities.

In the discrete domain, tokenised strokes bind physical execution to psychological orientation along the p0–p1 continuum (Fig. 4.5).

6.5 Extended Quantification (E15)

Whilst DQ4 provides a clear interpretive anchor, higher level analysis often requires finer resolution. The E15 system extends DQ4 by subdividing each attribute into fifteen levels, enabling more precise tracking of tactical escalation and degradation.

Each DQ4 code is mapped onto the E15 scale using

$$\varepsilon = 2^{i} - 1 \tag{6.1}$$

Table 6.4 Attribute strength in the E15 system

DQ4 code	Spin strength ε_O	Speed strength ε_I
0/1	$\varepsilon_O(0) = 1$	$\varepsilon_I(1) = 1$
00/11	$\varepsilon_O(00) = 3$	$\varepsilon_I(11) = 3$
000/111	$\varepsilon_O(000) = 7$	$\varepsilon_I(111) = 7$
0000/1111	$\varepsilon_O(0000) = 15$	$\varepsilon_I(1111) = 15$

where i is the number of digits in the corresponding DQ4 code. The four DQ4 levels therefore map to the set $\{1, 3, 7, 15\}$ on the E15 scale (Table 6.4).

This refinement enables tactically significant distinctions to be expressed explicitly. For example, the difference between a rally-building spin ($\varepsilon_O = 4$) and a point-ending spin ($\varepsilon_O = 7$), both of which may fall within the same DQ4 category, can be distinguished quantitatively within the E15 scale. Such distinctions are particularly useful for tracking tactical persistence: whether a player maintains their characteristic strength level ($\varepsilon_O, \varepsilon_I$) across exchanges or whether it degrades under pressure.

For comparison, strength values may be normalised according to

$$\tilde{\varepsilon} = \frac{\varepsilon}{15} \tag{6.2}$$

E15 values can always be re-binned into DQ4 categories when broader classification provides greater clarity. The extended scale therefore refines the analysis without replacing the discrete interpretive framework on which TTNS is built.

6.6 Fundamental Stroke Vocabulary

The discrete attributes and quantification systems are consolidated in Table 6.5, forming the core vocabulary of TTNS.

This dual quantification schema balances intuitive interpretation (via DQ4) with analytical precision (via E15), making the system suitable for modelling, annotation, training analysis and comparative studies.

Whilst such extended representations naturally support time-series analysis and algorithmic interpretation, a detailed treatment of their AI applications is beyond the scope of this work. The possibility is, however, indicated later in this chapter.

6.7 Discrete Stroke Geometry

This section defines the three 'ingredients' of a stroke: Hit, Brush, Contact.

1. **Hit directions**

- **Hit^{+1}**: forwards–upwards

Table 6.5 Characteristics and quantification of the fundamental attributes

Attribute	Characteristic	DQ4 (anchor codes)	E15 (extended scale)
Hit (F_h)	Flat, forwards, instinctive and impulsive.	1 = nudge	$\varepsilon_H \in \{1, 3, 7, 15\}$
		11 = punch	
		111 = smash	
		1111 = strike	
Brush (F_b)	Tangential, rotational, intentional and tentative.	0 = caress	$\varepsilon_B \in \{1, 3, 7, 15\}$
		00 = graze	
		000 = skim	
		0000 = sweep	
Speed (v)	Compress reaction window and apply pressure.	1 = slow	$\varepsilon_I \in \{1, 3, 7, 15\}$
		11 = fast	
		111 = very fast	
		1111 = exceptionally fast	
Spin (ω)	Inject high entropy and dictate rally tempo.	0 = light	$\varepsilon_O \in \{1, 3, 7, 15\}$
		00 = strong	
		000 = very strong	
		0000 = exceptionally strong	

- **Hit^{-1}**: forwards–downwards
- **Hit^{0}**: a perfectly horizontal hit although unrealistic.

Therefore, there are three possible hitting directions.

2. **Brush directions**

There are four possible directions:

- **Upwards**: Topspin (σ)
- **Downwards**: Backspin (∂)
- **Lateral**: Left-lateral (Lθ) and right-lateral (Rθ)

3. **Ball contact geometry**

Brushing may be applied to the rear of the ball in multiple directions, and the contact regions (i.e., shoulder, equator and back) are described in Sect. 2.11.1. However, a downward brush on the shoulder is unrealistic, as the motion is limited by biomechanical reach and bat orientation.

6.8 Stroke Space and Muscle Memory

The mechanically complete contact space of Sect. 4.3, under-contact, over-contact and lateral modulation, is inherently continuous. To reveal its structure in a form that can be practised and internalised, this space is now discretised into a finite 'stroke space'. Every primitive class of strokes (backspin dominant, topspin dominant, lateral compound) finds its representation within this discretised counting.

The ball's rear equator and back area are treated as one contact region, as this allows all four brushing directions (Fig. 2.9b). The shoulder region allows three brush directions: upwards and the two lateral directions.

1. **Impact on the rear equator and back**

All four brush directions are permitted: upwards, downwards, left-lateral and right-lateral.

- Without brush:
  15 hit levels × 1 hit direction (Hit^{+1} only) = 15 strokes.
- With brush:
  15 hit levels × 2 hit directions (Hit^{+1} and Hit^{-1}) × 15 brush levels × 4 brush directions = 1,800 strokes.
- Subtotal: 1,815 strokes.

2. **Impact on the shoulder**

Downward brush is excluded by geometry.

- Without brush:
  15 × 1 (Hit^{-1} only) = 15 strokes.
- With brush:
  15 hit levels × 2 hit directions × 15 brush levels × 3 brush directions = 1,350 strokes.
- Subtotal: 1,365 strokes.

Total theoretical stroke count: 3,180.

This count maps directly onto the primitives of Sect. 4.3. Under-contact (backspin) configurations arise predominantly from the rear equator region with downward brush; over-contact (topspin) from upward brushed rear impacts and many shoulder impacts; lateral brushing generates the compound forms (sidespin and two lateral fades). The 3,180 configurations therefore discretely sample the same complete possibility space defined earlier by the three fundamental classes.

Whilst this count is mathematically exact under the stated assumptions, it represents a discretised abstraction of an inherently continuous physical reality. In practice, hit direction, brush angle and force magnitude are not fully independent but vary along smooth, coupled spectra shaped by biomechanics and timing.

Discretisation does not fragment the game but makes its complexity intelligible. It explains why skill is developed through sustained, purposeful practice: players navigate a discrete space of 3,180 stroke configurations, selectively developing muscle memory for a manageable subset of high-reliability patterns. Configurations involving brushing are far more numerous than purely hitting-based strokes, which is one reason spin-based strokes are typically more refined, variable and technically advanced. Repeated interaction compresses explicit control into subconscious execution. In TTNS terms, a player's internal model is a personally optimised subset of these theoretical possibilities. That subset is precisely their practical realisation, or lack of mechanical completeness: a player who has fully explored the under-contact, over-contact and lateral regions will cover all three classes; one who neglects, e.g., over-contact off the backhand will exhibit the structural vulnerability anticipated in Sect. 4.3. Repetition is therefore the process by which an immense theoretical space distils into a robust tactical repertoire, the practical synthesis of the Hit–Brush dialectic into automatic, intentional play.

6.9 Summary: The Discrete Framework

This chapter establishes a discrete analytical framework for table tennis by quantising continuous physical quantities into four fundamental attributes: *Hit* (through-centre force), *Brush* (tangential force), *Speed* (translational velocity) and *Spin* (rotational velocity). These attributes form the mechanical primitives from which all strokes are constructed.

To operationalise this framework, two complementary quantisation systems are introduced. DQ4 provides tactically intuitive four-level categories for stroke execution (i.e. Nudge (1) $\rightarrow$ Punch (11) $\rightarrow$ Smash (111) $\rightarrow$ Strike (1111), in combination with, Caress (0) $\rightarrow$ Graze (00) $\rightarrow$ Skim (000) $\rightarrow$ Sweep (0000)), enabling practical classification of force application. E15 extends this structure to fifteen resolution levels for finer-grained analysis using the mapping $\varepsilon = 2^i - 1$, allowing scalable representation of stroke intensity (e.g., speed strength $\varepsilon_I = 8$, spin strength $\varepsilon_O = 5$) across a wider mechanical range.

Hybrid actions arise through binary combinations of hitting and brushing components (e.g. 0011 for slice), demonstrating how discrete primitives recombine to represent the diversity of real stroke techniques. When direction and contact geometry are incorporated, the framework yields a theoretical stroke space of approximately 3,180 mechanically distinct possibilities. In practice, effective play internalises only a small, individually optimised subset of this space through sustained training and tactical adaptation.

By transforming continuous mechanics into countable events, the framework establishes a precise vocabulary for describing stroke execution. This discretisation forms the analytical foundation for the notation and tactical modelling developed throughout the remainder of the discrete analytical framework.

6.10 AI System Integration

Building on the I–F–TC event grammar introduced in Sect. 2.9.2, the discretisation framework developed in this chapter converts physically continuous rally events into tokenised stroke and tactical states. By transforming continuous mechanical variables into discrete, tokenised attributes, the dynamics of play are recast into a finite, structured state space. This representation aligns directly with the requirements of artificial intelligence, where perception, decision-making and learning operate over discrete or discretised inputs (Russell and Norvig 2003).

At its core, the Table Tennis Notation System (TTNS) functions as a symbolic language. Each stroke is encoded as a sequence of discrete tokens, and each rally as a sequence of state transitions. This enables the game to be expressed not only as physical interaction but as a *computable process:*

$(J, S) \rightarrow E\ (v, \omega) \rightarrow T\ (E, G) \rightarrow$ response.

In this form, the rally becomes a structured progression of states that can be recorded, evaluated and predicted.

1. **Discretisation as a bridge to learning**

The DQ4 and E15 systems provide a natural interface between physical measurement and machine learning. In this chapter, continuous inputs (e.g. velocity, spin) may be:

- Quantised into discrete tokens for symbolic modelling
- Retained as continuous variables for statistical or neural network approaches

This dual compatibility allows the framework to support both:

- Model-based AI, where rule and mechanisms remain interpretable, and
- Data-driven AI, where patterns are learned from examples.

In reinforcement learning (RL), such discretisation is closely related to state aggregation and tile-coding, where continuous domains are organised into tractable representational units (Sutton and Barto 2014). TTNS therefore enables structured learning by bounding the space of possible actions and outcomes without severing the link to the underlying physics.

2. **State representation and decision processes**

Within TTNS, each rally can be represented as a sequence of tactical states (Sect. 4.6.5):

$T_n = (E_n, G_n) \rightarrow T_{n+1} = (E_{n+1}, G_{n+1})$

This structure is analogous to state-based decision frameworks used in artificial intelligence, including Markov decision processes. The current state encodes available information, actions transform the state, and policies determine action selection. In this context, the stroke token functions as an action, whilst rally evolution provides the transition structure.

Stroke selection becomes a decision process over a finite action space, guided by the objective of:

- Increasing technical pressure φ *(J, S)*, shaping tactical pressure *T (E, G)*.
- Managing entropy E_p (Table 6.2).
- Maintaining initiative.

Higher-level tactical intentions, such as initiating pressure, stabilising the rally or finishing the point, may also be viewed as temporally extended actions, analogous to options in hierarchical reinforcement learning (Sutton et al. 1999).

3. **Combinatorial structure and practical reduction**

The theoretical stroke space of approximately 3,180 combinations defines a finite but high-dimensional action domain. From an AI perspective, this space is:

- Sufficiently rich to model tactical diversity
- Sufficiently bounded to allow tractable exploration

As in human learning, effective strategies are developed not through exhaustive enumeration but through the identification of high-reliability subsets. This parallels reinforcement learning processes, where optimal policies are discovered through repeated interaction rather than explicit enumeration of all possible actions (Minh et al. 2015). The progressive development of a player's repertoire also resembles curriculum learning, in which simpler or more stable action patterns provide the foundation for later tactical complexity (Bengio et al. 2009).

4. **Interpretability and symbolic transparency**

A key advantage of TTNS lies in its interpretability. Each tokenised action retains a direct physical meaning:

- 1100 encodes a specific balance of Hit and Brush,
- 0011 encodes a different mechanical and tactical intent.

This transparency allows AI systems built upon TTNS to remain explainable:

- Decisions can be traced to mechanical causes.
- Tactical patterns can be analysed in human-readable form.
- Model outputs can be validated against physical principles.

In this respect, TTNS provides a symbolic backbone for explainable AI. Rather than treating strokes as opaque labels, it preserves the causal link between mechanics, tactical intention and decision outcome. This allows coaches, players and AI systems to exchange tactical knowledge through a shared, physically grounded vocabulary (Garce & Lamb 2020, Newell & Herbet 1976 and Lipton 2018).

5. **Notation readiness**

The framework developed in this chapter provides more than a descriptive language. It defines:

- A hierarchical state space $T\,(E, G) \leftarrow E\,(E_v, E_\omega)$ and $G\,(\alpha, z) \leftarrow U\,(v, \omega) \leftarrow$ (Hit, Brush)

- An action space (tokenised strokes)
- A transition structure (rally evolution $I_1 \rightarrow F_1 \rightarrow TC \rightarrow F_2 \rightarrow I_2$)

Together, these constitute the essential representational components of an intelligent system: state, action and transition. In machine learning, the design of representations is often one of the most consequential steps in determining what an agent can perceive, learn and generalise (Bengio et al. 2013).

TTNS is one such representation. It provides the alphabet and grammar through which table tennis can be expressed as a computable, interpretable and physically grounded decision system.

References

Bengio Y et al. (2009) Curriculum Learning: Proceedings of the 26th International Conference on Machine Learning. Montreal. https://www.Curriculumlearning

Bengio Y et al. (2013) Representation learning: A review and new perspectives. IEEE Transactions on Pattern Analysisand Machine Intelligence. 35(8), 1798–1828. https://arxiv.org/abs/pdf/1206.5538

Garcez A, and Lamb LC (2020) Neurosymbolic AI: The 3rd wave. https://arxiv.org/abs/2012.05876

Lipton ZC (2018) The mythos of model interpretability: Communication of ACM. 61(10), https://spawn-queue.acm.org/doi/10.1145/3236386.3241340

Mnih V et al. (2015) Human-level control through deep reinforcement learning: Nature 518. pp.529–533.https://doi.org/10.1038/nature14236

Newell A, and Herbert AS (1976) Computer science as empirical inquiry: symbols and search. Communications of theACM. 19(3), 113–126. https://doi.org/10.1145/1283920.1283930

Russell S, and Norvig P (2003) Artificial intelligence: A Modern Approach. Second Edition. https://www.newchap02.pdf

Sutton RS, and Barto AG (2014) Reinforcement Learning: An Introduction (2nd ed) Chapter 9. MIT Press.https://www.SuttonBartoIPRLBook2ndEd.pdf

Sutton RS, Precup D, and Singh S (1999) Between MDPs and Semi-MDPs: A framework for temporal abstraction in reinforcement learning. https://www.aij.dvi

Chapter 7
Quantifying Tactical Strategies

Table tennis unfolds under constant dialectical tension: the player must choose between direct force application and subtle disruption, between resolving the rally immediately and prolonging it to induce error. Whilst concrete tactical patterns such as the third-ball attack or the first offensive stroke (FOS) represent specific technical events, their underlying logic arises from more fundamental strategic oppositions.

Building on the dialectical framework established earlier, this chapter formulates and quantifies this core tactical dichotomy using the DQ4 and E15 scales. The transition moves analysis beyond static labels towards a dynamic, computable model of player strategy, in which each shot serves as a data point within a broader tactical argument.

7.1 Winner Shots and Error Forcing

Every tactical decision in table tennis implicitly answers a single question: should the current ball state be resolved immediately or destabilised further? The tension is expressed dialectically as a *tactical Thesis* (Tactic1) and *Antithesis* (Tactic0).

Tactic1: Winner Strategy (Resolution Operator)

Tactic1 represents the strategic mandate to resolve the rally by injecting sufficient energy and/or placing the ball into geometry that disadvantages the opponent, and terminating the point within the current interaction cycle. Tokenised as 1, Tactic1 embodies the p1 logic of order: it seeks to collapse rally uncertainty by overwhelming the opponent's reaction window (t_w) through speed, depth, angle and placement. Its success criterion is immediate non-return.

Tactic0: Error-Forcing Strategy (Destabilisation Operator)

TacticO represents the strategic mandate to prolong the rally whilst deliberately increasing informational and mechanical entropy until the opponent's internal model

X. Ji, *Spin, Speed and Strategy*, Mathematical Engineering,
https://doi.org/10.1007/978-3-032-25518-1_7

Table 7.1 Dialectical tactics and p0–p1 continuum

Hierarchy	Concept	Timescale
Structural	p0–p1 continuum	Long-term (style, identity)
Tactical	Tactic0 / Tactic1	Point-by-point
Technical	ε_H, ε_B, ε_1, ε_0	Stroke-by-stroke
Physical	F_h, F_b, J_h, J_b	Instantaneous

fails. Tokenised as 0, Tactic0 embodies the p0 logic of chaos: it withholds terminal force in favour of variation, persistence and uncertainty. Its success criterion is opponent error rather than direct resolution.

The balance between Tactic1 and Tactic0 is inherently fluid, shifting with the score and the player's internal execution state.

The evolution of a rally can be represented as a sequence of tactical actions applied alternately by the two players:

$$(E_n, G_n) \xrightarrow{T_{1/0}} (E_{n+1}, G_{n+1}) \tag{7.1}$$

where $T_{1/0}$ denotes the tactical operator applied at time n (discrete format of t), applying Tactic1 (T_1) or Tactic0 (T_0). The transition expresses the rally as a dynamic interaction in which each tactical decision modifying the evolving state of play.

The p0–p1 continuum, established in earlier chapters, characterises a player's preferred operating regime in terms of entropy tolerance and rally structure. Tactic0 and Tactic1 do not replace this continuum; they represent its point-by-point tactical realisation. Whilst the p0–p1 continuum describes where a player tends to stabilise over time, Tactic0 and Tactic1 describe how the player acts on a given ball state.

This multi-level relationship is summarised in Table 7.1, which situates tactical decisions within the hierarchy linking physical action, technical parameters and long-term playing identity.

Tactic1 resolves uncertainty through force; Tactic0 amplifies uncertainty until force becomes unnecessary.

7.2 Quantifying Tactical Profiles

The tactical proficiency of a player within each strategic logic is quantified through the ε-factor. These values (Table 7.2), derived from the discrete quantification methods of the DQ4 and E15 frameworks, provide a standardised measure of strategic effectiveness.

- ε_{T1} – strength for Tactic1 (winner strategy):

This parameter measures a player's ability to reliably convert favourable ball states into terminal outcomes.

Table 7.2 Fundamental tactical dichotomy: DQ4 anchors and E15 strengths

Attribute	Definition	Notation	DQ4 code	E15 strength
Tactic1 (winner)	Playing a decisive shot that the opponent fails to return	1	1 = present	$\varepsilon_{TI} \in \{1, 3, 7, 15\}$
			11 = habitual	
			111 = intentional	
			1111 = exceptional	
Tactic0 (error-forcing)	Inducing mistakes through variation and persistence	0	0 = present	$\varepsilon_{TO} \in \{1, 3, 7, 15\}$
			00 = habitual	
			000 = intentional	
			0000 = exceptional	

- ε_{TO} – strength for Tactic0 (error-forcing strategy):

This parameter measures a player's capacity to maintain rally stability whilst progressively increasing entropy within the opponent's internal model.

By mapping ε-values across rallies within a match, a player's *tactical profile* can be represented as a dynamic state rather than a fixed label. Changes in the relative balance between ε_{TI} and ε_{TO} reveal transition points where a player moves between resolution-dominant and destabilisation-dominant strategies, often in response to fluctuations in the opponent's internal entropy state.

Example profile

A player with ($\varepsilon_{TI} = 7$, $\varepsilon_{TO} = 3$) is *intentionally winner-oriented* with *habitual error-forcing capacity*. Such a profile corresponds to a classical p1-centred power player who must make deliberate tactical adjustments when engaging in extended disruptive exchanges, e.g., increasing $\varepsilon_{TO} = 3 \rightarrow \varepsilon_{TO} = 6$.

A *tactical profile* therefore describes *preference* and *reliability rather than obligation.* Players may execute either tactic but only within the bounds of their ε-capacity; beyond this range tactical intent exceeds technical and strategic feasibility.

7.3 The Technical–Strategic Chain

Tactical strengths (ε_{TI}, ε_{TO}) are not abstract constructs but the strategic synthesis of the core technical attributes (ε_H, ε_B, ε_I, ε_O). This relationship forms a direct and measurable causal chain (Table 7.3), illustrating how physical force and impulse propagate through technical execution to produce higher-level strategic outcomes.

A player's tactical proficiency is consequently limited by technical capability. An *exceptional* Tactic1 (winner strategy strength, e.g. $\varepsilon_{TI} = 7$) cannot be realised if the underlying hitting capacity ε_H remains at the *Punch* level ($\varepsilon_I = 3$). Within this framework, apparent strategic failure is often not a failure of intent but the downstream consequence of a technical bottleneck in the underlying stroke attributes.

Table 7.3 Tactical–strategic chain

Strategic mandate	Driven by (technical attributes)	Primary goal
Tactic1 (winner)	High ε_H (hit), high ε_I (speed)	Maximise E_v (translational kinetic energy), minimise t_w (reaction window)
Tactic0 (error-forcing)	High ε_B (brush), high ε_O (spin)	Vary ω (angular velocity), maximise entropy and uncertainty

The technical–strategic chain therefore provides the mechanical foundation upon which quantifiable tactical behaviour is constructed.

7.4 Tactical Failure in the Discretised Domain

The distinction between execution failure (fidelity loss) and model failure (validity loss) was developed in the continuous domain of physical mechanics (Sect. 4.6.4). Within the discretised domain of TTNS, this architectural distinction maps directly onto a vertical dependency chain, as presented in Table 7.4.

Tactical failure in the discretised domain frequently manifests as a cross-layer *category error*: the meta-model selects a tactical mandate ($\varepsilon_{TI} = 5$, $\varepsilon_{TO} = 4$) that exceeds the player's technical capacity ($\varepsilon_H = 3$, $\varepsilon_B = 3$). In this scenario, what appears to be *validity loss* (a strategic failure) is driven by an ungrounded *fidelity loss*; the failure is not a lack of tactical intent, but the physical infeasibility of the mandate itself.

Analytically, these three levels form a vertical dependency chain. Cognitively, however, they do not operate sequentially in real time. Through repeated rehearsal, the meta-model's layered architecture compresses. The player's coordination layer ceases to process physical force, technical capacity, and tactical intent as discrete, sequential steps. Instead, the bounded reflection mechanism relies on highly compiled, experienced-based triggers that bypass lower-level computation entirely. A trained player immediately perceives a 'winner opportunity' or registers a hard counter from experience without explicitly computing ε_H or ε_{TI}. The vertical chain collapses into rapid, intuitive action selection, allowing the meta-model to execute framework maintenance or shifts within millisecond-scale decision windows.

Table 7.4 Vertical dependency chain of tactical failure

Level	Domain	Component	Control failure
1	Physical	Force: (hit, brush)	Fidelity loss
2	Technical	ε measures: ε_H, ε_B, ε_I, ε_O	Fidelity loss
3	Tactical	T_1 (winner) vs T_0 (error-forcing), and ε measures: ε_{TI}, ε_{TO}	Validity loss

This formulation preserves the analytical distinction from Sect. 4.6.4 whilst grounding it explicitly in TTNS primitives and respecting the cognitive reality of millisecond-scale decision-making.

7.5 Comparative Frameworks and Practical Application

Other analytical frameworks provide useful descriptive overviews of performance. The widely used Hexagon Model (Fig 7.1), for example, organises player capability across six dimensions: Power, Speed, Tactics, Serve, Defence, and Mental Toughness. These indicators are practically meaningful and broadly representative of differences in performance level. **Ma Long**, widely regarded as one of the greatest players in the history of the sport, is frequently cited as the exemplar of this model: a player whose dominance reflects not a single exceptional attribute but the rare convergence of all six dimensions at the highest level simultaneously. His career embodies what complete, balanced excellence looks like in competitive play.

Whilst the Hexagon Model describes *what* a player can do, it does not explain *how* those capabilities interact during a rally, nor does it provide a mechanism for quantifying the moment-to-moment strategic tension of match play. By contrast, the (ε_{T1}, ε_{T0}) framework isolates the core dynamic of play directly: the continuous negotiation between Tactic1 and Tactic0 complexity. It translates tactical decisions into analysable, quantifiable data, moving from descriptive profiling towards a causal, physics-grounded account of why rallies unfold as they do.

Applications of the ε-based tactical framework include:

- Precision training

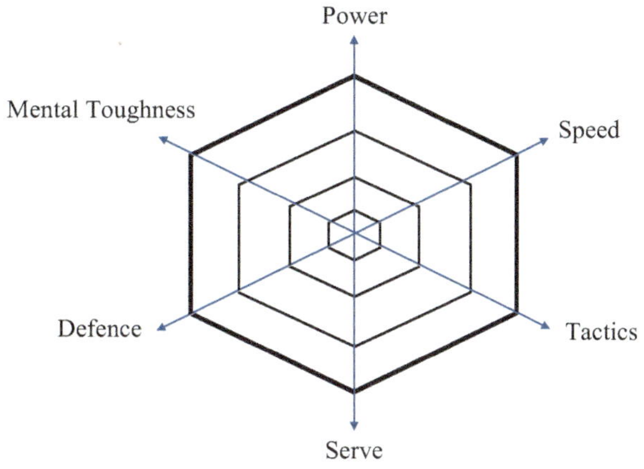

Fig. 7.1 Radar chart mapping six core attributes of athlete performance: serve, defence, mental toughness, power, speed and tactics

Distinguishes whether a tactical failure arises from poor execution (typically a technical bottleneck in ε_H or ε_B, or from an inappropriate strategic mandate such as a mismatch between ε_{TI} and ε_{TO}.

- In-game adaptation

Against a patient, structurally sound defender (high ε_{TO}), a player may need either to elevate ε_{TI} to an *exceptional* level to force resolution or to engage deliberately in a Tactic0 exchange to probe the opponent's tolerance for entropy.

- Opponent profiling

Opponent assessment becomes the estimation of an opponent's (ε_{TI}, ε_{TO}) profile, allowing prediction of the entropy regime they are likely to impose on the rally.

7.6 Tactical Dialectics

The Tactic1–Tactic0 framework derives its explanatory power from a dialectical tension: winner strategy (thesis) versus error-forcing strategy (antithesis). This tension renders table tennis analytically rich rather than tactically ambiguous. Recent match data from the autonomous robot Ace (Durr et al. 2026) illustrate that human players generally won points through stronger-than-average, high-speed and high-spin shots, consistent with Tactic1 and the direct resolution of the rally. By contrast, the speed distributions of Ace's winning and successfully returned shots were similar, indicating that its points were secured less through terminal force than through consistent return and the accumulation of pressure associated with Tactic0.

By assigning discrete ε-values (ε_{TI}, ε_{TO}), TTNS provides a unifying language for the physics of Chapters 2 and 3, the tactical mechanics of Chap. 4, the dialectical reasoning of Chap. 5, and the discrete formalism of Chap. 6. The sport is framed not as a collection of techniques but as a *dynamic system of managed contradictions*, resolved point by point through constrained tactical choice.

Reference

1. Dürr P, El Gheche M, Maeda GJ et al. Outplaying elite table tennis players with an autonomous robot. Nature 652, 886–891 (2026). https://doi.org/10.1038/s41586-026-10338-5

Chapter 8
Encoding the Essential Strokes

For decades, table tennis strokes have been catalogued using descriptive names and subjective nomenclature. Beneath this diversity lies a fundamental structural question: *what underlying logic connects strokes such as Push and Chop or Drive and Smash?*

The Table Tennis Notation System (TTNS) provides the answer. It is a universal stroke language grounded in the physics of contact and the strategic dialectics of match play established in the preceding chapters.

By treating every stroke as a digital combination of Hit and Brush and every motion as a combination of Speed and Spin, TTNS transforms the subjective feel of play into an objective, computable data structure.

TTNS serves two complementary purposes:

- **Binary stroke encoding**: Each stroke is represented by a compact binary code that directly reflects the fundamental contact forces: Hit (F_h) encoded as 1, and Brush (F_b) encoded as 0.
- **Structural stroke relationships**: The notation reveals how complex techniques (e.g. *slice*, Fig. 6.1) are constructed from combinations of simple mechanical components, and how these components align with the Tactic1 (Winner) and Tactic0 (Error-Forcing) strategic mandates.

TTNS does not replace familiar stroke techniques such as *loop, block or flick*. Rather, it augments them with a precise symbolic language that bridges coaching intuition, player execution and analytical rigour.

X. Ji, *Spin, Speed and Strategy*, Mathematical Engineering,
https://doi.org/10.1007/978-3-032-25518-1_8

8.1 Binary Coding

Traditional coaching systems catalogue large repertoires of strokes, often listing 20–30 variations per wing organised into descriptive labels and stylistic families. Although these taxonomies vary across coaching traditions, they rest on a shared mechanical foundation.

TTNS treats forehand and backhand not as fundamentally distinct categories but as symmetrical expressions of the same underlying mechanics. This reflects the reality of modern play, where two-winged competence is essential.

TTNS represents each stroke as a *binary sequence* encoding the ordering and dominance of Hit (1) and Brush (0) actions during the brief ball–bat contact interval.

The code $b_3b_2b_1b_0$ forms a scalable coding language.

- 2-bit codes generate four *basic strokes* (Fig. 8.1).
- 3-bit codes extend the vocabulary to *intermediate rally strokes* (Fig. 8.2).
- 4-bit codes describe *advanced strokes* (Table 8.3).

This structure is sufficiently compact to encode all established stroke techniques yet flexible enough to accommodate strokes not currently in the coaching terminology.

8.1.1 Basic Strokes (2-Bit)

All table tennis strokes decompose into two fundamental components: Hit (1) and Brush (0). The 2-bit code b_1b_0 represents their sequence during contact, forming the *atomic units* from which higher-level strokes are constructed.

Figure 8.1 illustrates how the four basic strokes are derived from the binary pairing.

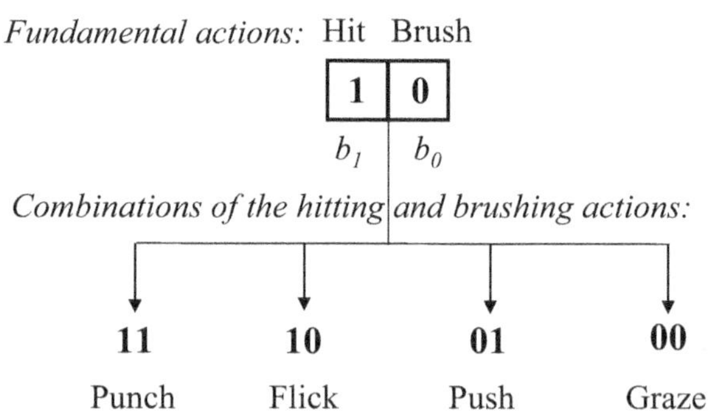

Fig. 8.1 Logical diagram of the 2-bit binary system for basic strokes, in which Bit 1 represents *Hit* and Bit 0 represents *Brush*, producing four primary stroke types

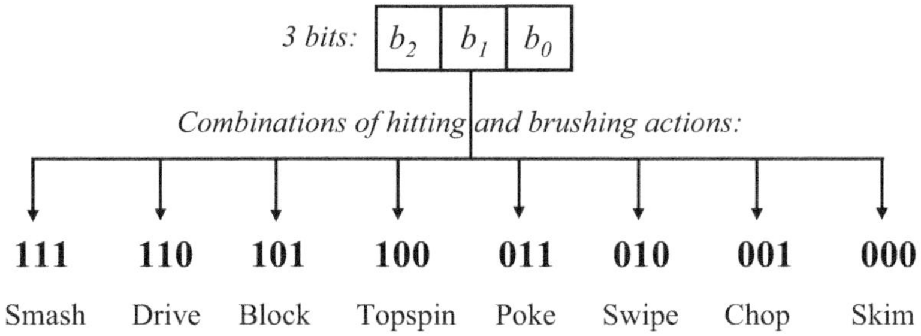

Fig. 8.2 Three-bit binary model (b_2, b_1, b_0) defining eight intermediate strokes through varying intensities of hitting and brushing actions

- **11 (Punch):** A firm, forward-driven impulse dominated by F_h. Tangential components are minimal. Variants include Punch^{+1} (a slightly upward counter-hit) or Punch^{-1} (a forward-downward punch-block).
- **10 (Flick):** A Hit-then-Brush action. A compact upward hit is followed by a brief tangential lift at the ball shoulder, producing a light topspin flick (1σ).
- **01 (Push):** A Brush-then-Hit action. Controlled downward tangential contact on the back of the ball is stabilised by a soft hit, producing a controlled backspin push (ə1).
- **00** (**Graze or Sway**): A stroke with negligible effective normal-force component (low-F_h). A soft graze produces light spin or float, creating a deceptive low-energy contact.

The Graze/Sway stroke (coded 00) exemplifies how TTNS formal notation captures multi-facet spin states that natural language struggles to describe. Its code reflects a purely Brush-dominant action (F_b) with negligible Hit component ($\varepsilon_I \approx 0$), producing light spin or float.

Within TTNS, the compound spin tokens: *θσ, θə, θθ, σθ, əθ,* $\chi\chi$ encode the stroke's deceptive variability. Sway can generate topspin, backspin, sidespin, or no-spin ($\chi\chi$) from a near-identical contact, making it highly disruptive. Its slow speed becomes an asset: the ball lands unexpectedly short, low, or with an unpredictable bounce, forcing the opponent to hesitate, adapt and generate their own pace. Thus, Sway sacrifices speed to gain temporal and spatial control through unpredictability, a tactical logic now formally encoded in the notation 00 and its associated spin tokens.

These 2-bit strokes also underpin the *basic serve actions*, examined in the following chapter. Their mechanical characteristics, visual tokens and ε-trait profiles are summarised in Table 8.1.

Table 8.1 Basic strokes (2-bit TTNS coding)

Stroke	Code	Visual token	Contact / Mechanics	Key characteristic (ε traits)
Punch	11	11^{+1} or 11^{-1}	Direct, firm hit with dominant J_h	Assertive offensive intent; speed-focused ($\varepsilon_I = 3, \varepsilon_O \approx 0$).
Flick	10	1σ	Hit then brush; upward tangential lift	Sudden, surprising; peak-contact action ($\varepsilon_I = 2, \varepsilon_O = 1$).
Push	01	$\partial 1$	Brush then hit: Downward tangential contact	Controlled backspin; consistency focused ($\varepsilon_I = 1, \varepsilon_O = 2$).
Graze / sway	00	$\theta\sigma, \theta\partial, \theta\theta, \sigma\theta, \partial\theta, \chi\chi$	Soft graze with F_b dominant	Deceptive, low-energy spin or float ($\varepsilon_I \approx 0, \varepsilon_O = 3$).

8.1.2 *Intermediate Strokes (3-Bit)*

Introducing a third bit expands the coding system to *eight intermediate strokes* commonly used in rallies, as illustrated in Fig. 8.2.

Selected strokes from Table 8.2 are detailed below; the remainder follow directly from the established definitions.

- **010 (Swipe)**: An equatorial hit combined with tangential brushing, generating sidespin-dominated trajectories, including pure sidespin as well as sidespin–topspin and sidespin–backspin variations.
- **110 (Drive)**: Two forward hitting digits followed by a light upward brush. The drive introduces deliberate topspin whilst preserving speed, linearity and directional stability.
- **011 (Poke)**: A brushing action followed by a firm rear contact, producing sharp backspin in a fast jab-cut motion.
- **101 (Block)**: A stabilising hit near the ball's shoulder combined with spin modulation. The final '1' represents momentum supplied by the table rebound rather than active force generation.
- **111 (Smash)**: A maximal, height-dependent extension of the Punch. Impact occurs in the upper-rear quadrant of the ball, producing a strongly downward trajectory. The vertical sign is fixed as Smash^{-1}, whilst Smash^{+1} is mechanically undefined. The stroke converts height and timing margin directly into terminal speed.

Table 8.2 Intermediate strokes (3-bit coding)

Stroke	Code	Visual	Contact / Mechanics	Key characteristic (ε traits)
Smash	111	111^{-1}	Pure forward hitting: Maximal impulse J_h	Power hitting ($\varepsilon_I = 7, \varepsilon_O \approx 0$); decisive flat speed.
Drive	110	11σ	Forward hit with a light upward brush	Speed with stability ($\varepsilon_I = 6, \varepsilon_O = 1$).
Block	101	$1\sigma1$	Early contact (τ_{rising}); rebound-assisted redirection	Redirect opponent's strength ($\varepsilon_I = 5$, $\varepsilon_O = 2$).
Slow topspin	100	$1\sigma\sigma$	Soft hit followed by a strong graze	Controlled looping ($\varepsilon_I = 4, \varepsilon_O = 3$).
Poke	011	$ə11$	Brush followed by a downward punch	Fast backspin jab or cut ($\varepsilon_I = 3, \varepsilon_O =$ 4).
Swipe	010	$\theta1\theta$	Equatorial hit with integrated side brush	Sidespin-dominated stroke ($\varepsilon_I = 2, \varepsilon_O$ $= 5$). Five variants: Pure sidespin, sidespin with topspin, sidespin with backspin, topspin with right sidespin topspin with left sidespin
Topspin fade		$\theta1\sigma$	Sidespin combined with upward brush	
Backspin fade		$\theta1ə$	Sidespin combined with downward brush	
Banna Flick		$\sigma1R\theta$	Upward brush at the ball's outside shoulder	
Strawberry Flick		$\sigma1L\theta$	Upward brush at the ball's inside shoulder	
Chop	001	$əə1$	Strong downward graze followed by soft hit	Heavy backspin or side-backspin (ε_I $= 1, \varepsilon_O = 6$)
Slow slice		$ə\theta1$	Downward skim followed by soft hit; prolonged tangential contact	
Skim/float	000	$\sigma\theta\sigma$, $ə\thetaə$, $\chi\chi\chi$	Pure brushing with no effective hit	Maximum brushing strength ($\varepsilon_I \approx 0$, $\varepsilon_O = 7$); deceptive touch.

8.1.3 TTNS for Emerging Strokes

When professional players introduce new strokes, these techniques are typically named descriptively within the coaching community. TTNS accommodates such innovations by encoding strokes according to their force structure and contact logic rather than relying on descriptive terminology.

The *banana flick*, *strawberry flick* (Table 8.2) and *boomerang flick* (Table 8.3) belong to a family of backhand topspin–sidespin flicks, distinguished primarily by their contact geometry and relative balance of Hit and Brush actions:

- **Banana flick**: Backhand contact on the *outside shoulder* of the ball, producing topspin combined with sidespin; encoded as $\sigma 1\mathrm{R}\theta$.
- **Strawberry flick**: Backhand contact on the *inside shoulder* of the ball, producing topspin with the opposite sidespin orientation; encoded as $\sigma 1\mathrm{L}\theta$.
- **Boomerang flick**: A stronger, faster variant of the *Swipe family*, best represented by a 4-bit encoding such as $\sigma 11\theta$, reflecting increased *Hit dominance* layered onto a sidespin-dominated contact.

TTNS therefore captures emerging techniques by revealing the shared biomechanics underlying different naming conventions, allowing new strokes to be formalised without continually expanding the descriptive vocabulary of the sport.

8.1.4 Four-Bit Coding Rules

A 4-bit code $b_3b_2b_1b_0$ encodes the *ordering* and relative *dominance* of Hit and Brush actions within an advanced stroke. Although sixteen binary patterns can be generated, only a subset correspond to physically meaningful actions. Three rules govern the validity of a 4-bit TTNS code.

1. Contiguous-Hit rule

A genuine Hit represents a single continuous impulse. Consequently, all '1' digits must appear as a single contiguous block, reflecting the uninterrupted application of normal force.

2. Table-momentum exception

In block-type strokes, a trailing '1' represents momentum supplied by the table rebound rather than an active hit generated by the player.

3. Contact geometry rule

The relative ordering of *Hit* and *Brush* determines the resulting spin type:

- Hit before Brush → topspin (contact near the shoulder of the ball)
- Brush before Hit → backspin (contact in the lower rear quadrant)
- Symmetric binary patterns → sidespin (equatorial contact)

Table 8.3 Advanced strokes (4-bit coding)

Stroke	Code	Visual token	Key characteristics (ε traits)
Strike	1111	1111^{-1}, 1111^{0}	Explosive slap producing maximal impulse ($\varepsilon_I = 15$, $\varepsilon_O \approx 0$). Delivers overwhelming F_h through compact, directional authority; largely spin insensitive.
Whip	1110	111σ	Snappy acceleration with topspin emphasis ($\varepsilon_I = 14$, $\varepsilon_O = 1$).
Topspin (loop)	1100	$11\sigma\sigma$	Faster and more forceful than 100; explosive topspin loop ($\varepsilon_I = 12$, $\varepsilon_O = 3$).
Drive block	1001	$1\sigma\chi 1$	Counterstroke using the opponent's pace; returns incoming topspin with stabilised but weak counter-topspin. Incoming ball typically ($\varepsilon_I = 9$, $\varepsilon_O = 6$).
Punch block		$1\theta\chi 1$	Uses sidespin deflection for angle-changing counterplay.
Chop block		$1ə\chi 1$	Cuts under the ball at the bounce, producing sharp backspin.
Drop	1000	$1\chi\chi\sigma$	Soft placement, minimal active hit; passive reaction to incoming energy ($\varepsilon_I = 8$, $\varepsilon_O = 7$), producing residual topspin.
Lob	0111	$\chi 111$	High balls, defensive retrieval; Long flight weakens spin, residual spin may remain ($\varepsilon_I = 7$, $\varepsilon_O = 8$).
Fast swipe	0110	$\theta 11\theta$	Stronger and faster variant of the 010 family ($\varepsilon_I = 6$, $\varepsilon_O = 9$). Variants include strong sidespin, side-topspin and side-backspin. Fast topspin fade: Sidespin combined with upward brushing action. Fast backspin fade: Sidespin combined with downward brushing action. Boomerang Flick: Powerful sidespin–topspin flick from the backhand side.
Fast topspin fade		$\theta 11\sigma$	
Fast backspin fade		$\theta 11ə$	
Boomerang Flick		$\sigma 11\theta$	
Slice	0011	$\theta ə 11$	Heavy backhand slice producing strong backspin with a sidespin component ($\varepsilon_I = 3$, $\varepsilon_O = 12$).
Wiggle	0001	$\theta əə 1$	Complex side-backspin contact with wobbling trajectory ($\varepsilon_I = 1$, $\varepsilon_O = 14$).
Sweep	0000	$\sigma\sigma\sigma\sigma$, $əəəə$, $\theta\theta\theta\theta$, $\chi\chi\chi\chi$	Exceptional spin-dominant disruption stroke ($\varepsilon_I \approx 0$, $\varepsilon_O = 15$).

8.1.5 Advanced Strokes (4-Bit)

Extending the representation to four bits enables the encoding of advanced strokes with greater force complexity and finer tactical resolution (Table 8.4).

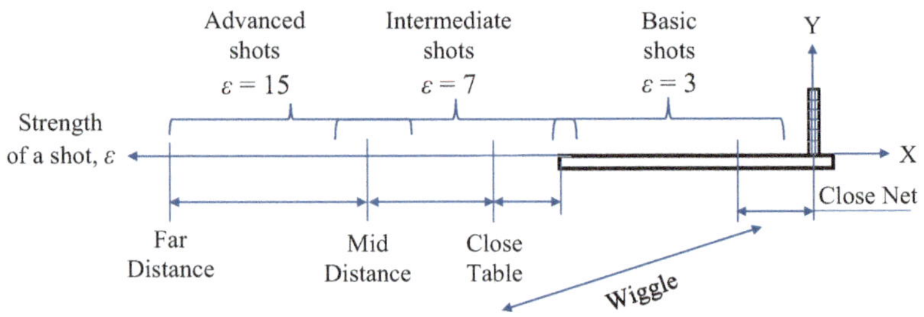

Fig. 8.3 Diagram relating a player's distance from the table to the mechanical strength of the shot being played. Shot strength values $\varepsilon = 3, 7, 15$ corresponds to close-table, mid-distance, and far-distance play, respectively

8.2 Strength Scaling (ε), Smash (111) and Strike (1111)

8.2.1 ε-Strength and Stance

A stroke's binary code defines its structural form, but its practical effect is scaled by the strength factor ε on the E15 scale. This relationship connects abstract notation to biomechanical reality. As illustrated in Fig. 8.3, a player's stance, particularly their distance from the table, correlates strongly with the ε-strength of an intended stroke.

8.2.2 Distinguishing Smash (111) from Strike (1111)

The Smash remains partially governed by the incoming spin state, whereas the Strike applies sufficient impulse to override it. This difference follows directly from the discrete quantification scheme.

1. **Quantitative separation (magnitude)**

From the E15 mapping,

- Smash (111) $\rightarrow \varepsilon_I = 7$
- Strike (1111) $\rightarrow \varepsilon_I = 15$

The transition from $\varepsilon_I = 7$ to $\varepsilon_I = 15$ marks a critical threshold: from an impulse that remains influenced by spin to one capable of dominating the incoming rotational state. Strike 1111 therefore encodes not just greater power but *power* sufficient to override the ball's existing spin.

2. **Spin sensitivity as the primary structural difference**

The defining distinction is spin sensitivity.

- **Smash (111)** is spin sensitive. Its outcome depends strongly on the incoming spin state: against backspin it drives into the net; against topspin it launches high; against sidespin the trajectory deflects laterally.
- **Strike (1111)** is spin insensitive. The applied normal impulse overrides the incoming spin, imposing a new translational trajectory that reflects player intent rather than inherited rotation.

3. **Mechanical interpretation**

- **Smash (111)** is a *geometry- and spin-conditioned terminal stroke*, remaining coupled to incoming angular momentum.
- **Strike (1111)** is an *impulse-dominant stroke.* The magnitude of F_h or J_h resets the relationship between translational velocity and spin; incoming angular velocity ω becomes a secondary perturbation.

4. **Height dependence (secondary but complementary)**

- **Smash (111)** remains *height-dependent*, relying on downward geometry (111^{-1}) that allows partial mitigation of the incoming spin.
- **Strike (1111)** is largely *height-independent* because impulse dominance neutralises rotational influence, allowing flatter or nearer equator contact (1111^{-1} or 1111^{0}).

5. **Tactical consequence**

A Smash (111) is a spin-sensitive, geometry-assisted terminal stroke whose success depends on alignment with the incoming spin state. A Strike (1111) is an impulse-dominant stroke that overrides rotational effects regardless of moderate spin.

The mechanical and tactical contrasts between these strokes are summarised in Table 8.4.

This difference reveals, rather than imposes, the *operational* distinction between Smash and Strike, allowing each to be developed as a distinct technique.

Table 8.4 Smash (111) versus Strike (1111)

Aspect	Smash (111)	Strike (1111)
ε_l strength	7	15
Dependence on ball height	High	Low
Dependence on downward angle	Essential	Incidental
Contact geometry	Shoulder, downwards	Shoulder or near-equator
Margin source	Height and timing window	Impulse magnitude and certainty
Failure mode	Loses dominance if ball is low	Retains dominance even on flat balls

8.2.3 Spin Neutralisation and Tactical Diversity

Section 3.4.2 established spin as a disruptor, a mechanism that reshapes flight, modifies bounce and challenges perceptual expectation. It also identified the counters: impulse override, rotational redirection and frictional decoupling. Within the continuous domain, these were framed as tactical choices. In the discrete TTNS framework, the same counters resolve into exactly two neutralisation pathways, not a spectrum of options, but a binary partition of mechanism.

Pathway 1: Impulse dominance

When the applied normal impulse J_h is sufficiently large, the incoming rotational momentum $I_A \cdot \omega$ becomes comparatively negligible. This is the regime of Strike (1111): translational momentum imposed by the player overwhelms the inherited spin state. The ball's trajectory reflects player intent, not incoming rotation. Spin is not countered; it is rendered irrelevant.

Pathway 2: Interactional disruption

A second pathway arises from *equipment-mediated interaction.* Rubbers such as long pimples or anti-spin surfaces surpress the frictional coupling between the bat and the ball, weakening or reversing the incident spin profile at contact. Where Pathway 1 outmuscles the spin, Pathway 2 un-couples it. The incoming rotation is not overcome by force; it is denied a mechanical foothold.

Convergent outcome, divergent mechanism

Both pathways produce the same discrete outcome: the opponent's spin ceases to be a hard counter. One pathway achieves this through force dominance (high J_h); the other through frictional decoupling (low μ). The result is identical. In continuous play, the distinction between these strategies is a matter of degree and timing. In the discrete domain, it is categorical.

Structural consequence: stylistic diversity

This binary partition has a structural consequence that extends beyond the shot level. Unlike many racket sports, where equipment behaviour is relatively uniform and effective technique converges towards a narrow mechanical ideal, table tennis sustains multiple viable strategic regimes precisely because both neutralisation pathways are competitively viable. Players may prevail either through:

- **Force dominance**: imposing decisive impulse and speed (e.g., Strike, Smash), or
- **Entropy injection**: manipulating spin, rhythm and surface interaction to destabilise the opponent's internal model of the rally.

Strategy, at the discrete level, is therefore organised around a fundamental binary: dominate the ball state or destabilise it. The two neutralisation pathways are not merely physical mechanisms; they form structural pillars in the strategic architecture of table tennis.

8.2.4 Equipment Diversity Innovation and the Direction of the Sport

The physical and tactical analysis developed in these chapters also raises a forward-looking question: where should equipment innovation end, and regulation begin? Table tennis has long accommodated substantial material diversity, including inverted rubbers, pips-out rubbers, anti-spin surfaces and no-sponge hardbats. This diversity has produced different mechanical regimes and tactical identities. It has also prompted recurring debate over whether extreme equipment variation should be restricted, or whether some forms of play should occupy separate competitive categories, as in sandbat competitions where the wooden blade is covered with sandpaper on each face.

The issue is not simply whether unusual surfaces are fair, but how far material architecture may legitimately shape the ball–bat interaction. If pips-out rubbers are valid because their surface geometry modifies friction, dwell time and spin response, then other geometries are conceptually possible. Straight, narrow grooves or channels, for example, might act by analogy with vehicle tyres, directing contact, increasing traction and assisting spin generation. Likewise, new surface materials may be designed to prolong dwell time, grip the ball more actively and release it with enhanced spin through material recovery rather than through the player's own impulse alone.

These possibilities reveal the regulatory boundary problem. A rubber surface that modifies friction through passive material structure remains within the traditional logic of equipment design. A surface that stores and releases energy more efficiently occupies a more ambiguous region but still belongs to material mechanics. Embedded sensors, now technically feasible, shift the issue again: if they measure impact force, incident velocity, exit velocity or contact location for post-match analysis, they may not directly alter the stroke. Micro-actuators, by contrast, would cross a different threshold. If integrated into the bat to assist speed or spin generation during impact, they would no longer be passive material properties but active mechanical intervention.

The central distinction is therefore between passive material response and active performance assistance. Regulation may reasonably permit diversity in friction, compliance, surface geometry and energy restitution, provided these arise from passive material properties and remain within authorised limits. It would need to be far more restrictive where electronics or actuators begin to supply information, force or spin in real time. The future direction of the sport may therefore depend on preserving tactical diversity while preventing equipment from replacing the player's perceptual and motor contribution.

In this sense, the long-term question is not whether table tennis should eliminate equipment variation, but how it should regulate innovation so that different mechanical regimes can coexist without turning the bat into an active device. The richness of the sport lies in the contest between human skill and material possibility; regulation must preserve both.

8.3 The Topspin Family

Under TTNS, traditional stroke families acquire a clear internal structure. When mapped onto the two-dimensional spin–speed plane (Fig. 8.4), the topspin family appears not as a linear escalation but as a *structured migration* across the coupled spin and speed.

The progression

The sequence from slow topspin (100) → flick (10) → (110) → (1100) → whip (1110) traces a deliberate rebalancing between rotational dominance and translational velocity.

- **Early variants** (slow topspin, flick) occupy the upper-left region: high spin at modest speed. These strokes are intrinsically disruptive: high arc, pronounced dip and bounce sensitivity whilst preserving temporal slack. Their tactical function is to destabilise rhythm and accumulate informational advantage, not to force immediate resolution.
- **The drive (110)** marks a decisive lateral shift: rotational richness is exchanged for speed, flattening the trajectory and compressing the opponent's reaction window whilst retaining topspin control.
- **The topspin (1100)** briefly restores rotational dominance at higher pace, occupying a central region of balanced spin–speed coupling.
- **The whip (1110)** completes the migration, prioritising speed and impulse. Located low and far to the right, it imposes maximal temporal pressure, leaving minimal

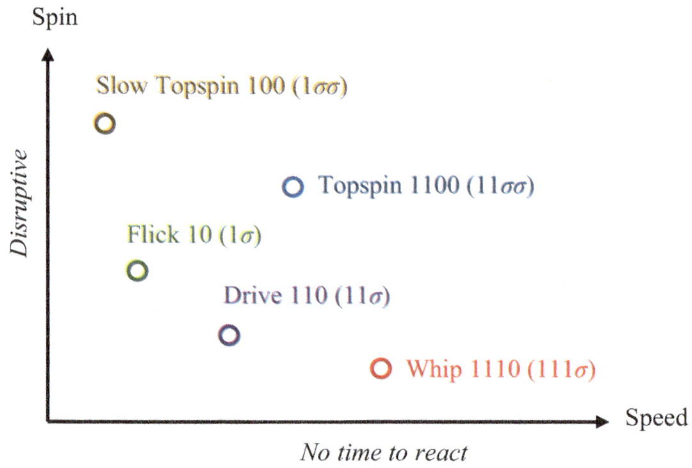

Fig. 8.4 Two-dimensional spin–speed diagram for topspin strokes. Spin increases vertically and speed increases horizontally, whilst the reaction window decreases at higher speeds. The plotted points indicate slow topspin (100) at high spin and low speed, flick (10), drive (110), topspin (1100) with balanced spin and speed, and whip (1110) at high speed and reduced spin

reaction time and sharply reducing tactical ambiguity. Here, disruption arises not from spin complexity but from time denial.

The reframing

TTNS therefore reframes topspin as an integrated, multi-modal strategic continuum, unified by intent (topspin) but differentiated by position within the spin–speed plane.

8.4 The p00–p11 States

This section discretises continuous p0–p1 continuum (Sects. 4.10) into four binary states along two dimensions:

- **Structural preference**: order (p1) or chaos (p0)
- **Regulation:** instinctive versus intentional (degree of conscious regulation)

The four codes represent *stable attractors* within the p0–p1 landscape.

8.4.1 p00: Unstructured Chaos (Instinctive p0)

A casual 'soft player' who disrupts stronger opponents through unconventional spin, timing disruption and deceptive contact, not by design, but by irregular interaction. In this sense, p00 represents the unstructured counterpart of the physics-level entropy mechanism described in Sect. 4.10.7: cue, spin and outcome are partially decoupled, but the disruption arises incidentally rather than through a controlled tactical model.

Features include:

- Wrist-dominated contact, producing weakly coupled or compound spin that are difficult to infer from stroke kinematics.
- Reduced forward acceleration, keeping the ball short or slow and removing the opponent's ability to utilise incoming pace for advanced attacking techniques.
- Equipment-induced irregularity (anti-spin, long pimples, or worn/slick inverted rubber), further decoupling input from output and generating erratic ball behaviour.

This phenomenon is often referred to informally as the 'club player's nightmare': the casual soft player who breaks the standard rules of timing and spin expectation. At higher levels, the effect is transient. Experienced players adapt by simplifying interaction, reducing stroke amplitude, stabilising timing, and prioritising push and placement over high-energy attacking strokes.

The casual soft player corresponds closely to the p00 profile, representing pre-structured chaos.

Characteristics

- Reliance on float balls, passive blocks and improvised pushes.
- Inconsistency in contact mechanics, with inconsistent mapping between applied impulse and resulting spin.
- Variation is incidental rather than through deliberately structured disruption.
- Entropy is introduced but without direction, calibration or persistence.

Instinct–intention axis

Predominantly instinctive. Internal entropy model is unstable; action is reactive, not anticipatory.

Tactical consequence

Technically stronger players may be unsettled locally by style ambiguity and timing irregularity, but not systematically. Under pressure, insufficient control degrades into inconsistency.

Interpretation

p00 shares superficial features with high-entropy play but lacks the governing structure required for sustained tactical advantage. Most beginners and unstable intermediates pass through this state.

Critical transition

A more fundamental reality is the phenomenon of *style mismatch* (or style clash), examined in Sect. 4.10.7. This further arises in two primary contexts: (1) the opening phase of a match, as players probe and calibrate, and (2) any encounter between unfamiliar opponents, including, at the highest level, a rising superstar (e.g. ages 18–20) who captures major titles as a relative unknown.

In both cases, the early phase of play creates an *exploratory regime*, an unmapped interaction domain where spin responses, timing windows and geometric preferences are untested. Within this phase, low-structure disruption, resembling the behaviour of a soft, irregular player, is functionally advantageous: ambiguity in spin, speed and timing delays the formation of a stable playing pattern and prevents rapid stabilisation of rally dynamics. Uncertainty is used to extend the opponent's calibration phase.

However, this advantage is transient. With exposure, opponents encode unfamiliar patterns. The player must complete transition from unstructured or novelty-driven disruption to stable, repeatable control. Competitive persistence depends not on initial disruption, but on the capacity to *stabilise and impose one's own game.*

8.4.2 p01: Structured Chaos (Intentional p0)

The trained disruptor. Chaos is deliberate, not accidental: controlled variation in spin, depth and tempo; systematic use of dead balls, sidespin and rhythm interruption; and a clear preference for error-forcing (Tactic0). Every variable is a weapon. Opponents collapse under decision overload, with no terminal force required.

8.4.3 p10: Controlled Order (Intentional p1)

The *trained attacker*. Order is imposed deliberately: repeatable topspin patterns, reliable first offensive shot (FOS) execution, and selective use of variation. Force is scheduled rather than impulsive. Rallies compress into predictable, high-probability exchanges in which technical superiority determines the outcome.

8.4.4 p11: Overdetermined Order (Instinctive p1)

Maximal commitment to force without sufficient adaptive modulation: heavy reliance on speed and power, minimal tolerance for disruption, and a persistent bias towards Tactic1 even in unfavourable ball states. Predominantly instinctive despite high technical capacity, with intention degrading as external entropy rises. Highly effective against passive opponents, yet structurally fragile against structured chaos (p01); an over-specialised state rather than a balanced game model.

8.4.5 Development from p00 to p11

Building on the continuous p0–p1 spectrum introduced in Fig. 4.5, Fig. 8.5 converts the same developmental logic into a discrete operational map. The purpose is not to repeat the continuum, but to distinguish structural preference from regulatory maturity: whether a player's order or chaos is instinctive, fragile and reactive, or intentional, stable and adaptable. The four resulting states (Fig. 8.5) are:

- p00-unstructured chaos: instinctive and reactive variation without control
- p01-structured chaos: intentional disruption, regulated p0
- p10-controlled order: intentional stability, regulated p1
- p11-rigid order: over-regulated execution, vulnerable to disruption

p01 and p10 are the mature states identified in Sect. 4.10.4: players capable of both generating and regulating entropy. By contrast, p00 and p11 are structurally fragile, lacking adaptability through insufficient control (p00) or excessive rigidity (p11). Development is therefore not a linear shift from p0 to p1, but a transition from instinct-driven behaviour to intentional regulation, captured by the pathways p00 → p01 and p11 → p10. Mastery resides in the capacity to stabilise, regulate and switch between regimes with intention.

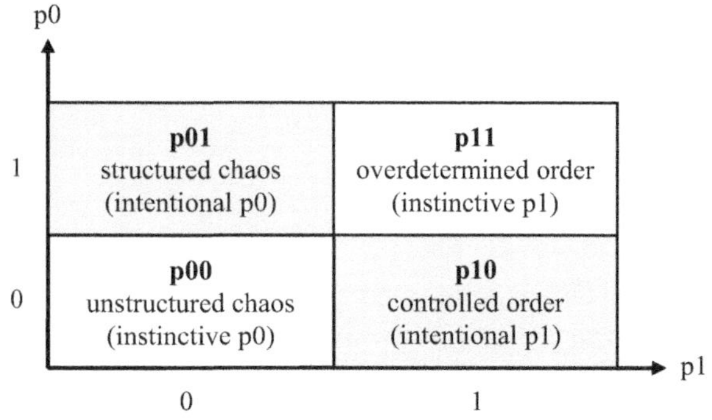

Fig. 8.5 The p00–p11 operational map, derived from the p0–p1 spectrum in Fig. 4.5. The continuum is discretised into four regulatory states according to structural preference and degree of intentional control. The stable competitive attractors are p01, structured chaos and p10, controlled order; p00 and p11 represent fragile states characterised respectively by insufficient control and excessive rigidity

8.4.6 Synthesis with Tactic0 and Tactic1

The relationship between structure, state and action is now explicit:

- **p0 / p1** → structural gravity
- **p00–p11** → operational state
- **Tactic0 / Tactic1** → moment-by-moment tactical control action

A p10 player may deploy Tactic0: a p01 player may seize Tactic1. What distinguishes them is not capability but where their system stabilises under uncertainty. What appears fluid in play is discrete in organisation, and it is this structure that enables TTNS to function as an analytical framework.

8.5 TTNS in Action

TTNS is both *physically grounded* and *structurally generative*. Each binary digit maps directly onto a specific force component applied during ball–bat contact:

- **1** encodes a hitting force (F_h) → producing speed strength (ε_1) → associated primarily with Tactic1.
- **0** encodes a brushing force (F_b) → producing spin strength (ε_0) → associated primarily with Tactic0.

A TTNS code therefore reveals the internal force sequence of a stroke. For example:

- **110 (Drive):** Hit–Hit–Brush sequence. Forward impulse dominates, whilst the final brushing component introduces upward rotational bias (σ).

- **011 (Poke):** Brush–Hit–Hit sequence. A downward brushing action (ə) structures the trajectory before punch-like hitting completes the impulse.

In each case, the code exposes the *mechanical ordering* of forces during contact. The notation functions as a representation of the stroke's tactical DNA.

Generative power

The generative character of TTNS is clearest in the 010 (Swipe) family. A single structural code represents dominant tangential brushing with minimal normal impulse. From this shared structure, multiple spin outcomes arise: pure sidespin, side-topspin or side-backspin, depending on the orientation of the brushing vector. TTNS thus separates *tactical intent*, the introduction of lateral disruption, from *technical realisation* the precise spin axis produced. This separation grants analytical power unavailable in traditional stroke naming, which often conflates appearance, outcome and intent.

Summary: from primitives to complexity

Just as spin symbols (σ, ə, θ, χ) condense rotational complexity, the binary structure of TTNS condense *stroke mechanics and tactical orientation* into concise codes. Compound spins are formed through combinations of spin symbols; analogously, compound strokes and tactical sequences are constructed from combinations of binary digits.

The consequence is foundational: table tennis technique can be described through a small number of mechanical primitives organised in binary form. From these primitives, the full diversity of strokes, variations and tactical patterns is generated. TTNS therefore illustrates a central principle of the game: complex play arises from simple physical interactions arranged in ordered sequences of hit and brush.

The binary representation introduced here describes the internal mechanics of individual strokes. In isolation, each code reveals the ordered dominance of hitting and brushing forces during contact. But strokes rarely occur independently. They combine into sequences that shape initiative, variation and tactical pressure across the rally, a transition from local mechanics to global match strategy.

Chapter 9
Encoding the Essential Serves

TTNS reframes the serve as a formal structure of mechanics, spin, placement and tactical intent. In doing so, it transforms serving from a tradition-bound art into a structured, analysable and optimisable system. The serve becomes not merely an opening action but a designed intervention into the future entropy state of the rally.

9.1 The Serve in TTNS

The serve is uniquely suited to TTNS analysis. It is the only stroke executed without responding to an incoming ball, the purest expression of intention within the game. All primary variables (Hit and Brush composition, speed and spin strength, placement and disguise) are selected rather than imposed, making each binary code a direct record of conscious mechanical choice and anticipated tactical outcome.

The serve therefore occupies a privileged position within the TTNS framework. Where rally strokes demand reactive adaptation, the serve represents premeditated strategy in its most distilled mechanical form, an ideal test case for demonstrating how discrete notation connects physical execution to tactical design.

9.2 Encoding the Serve Arsenal

Serves use the same binary library defined in Chap. 8. This continuity is a central strength: no special rules are required. A 01 (Push) serve and a 01 (Push) rally stroke share identical mechanical DNA, differing only in strategic context and temporal intent, not in force composition.

X. Ji, *Spin, Speed and Strategy*, Mathematical Engineering,
https://doi.org/10.1007/978-3-032-25518-1_9

9.2.1 Basic Serves (2-Bit)

Two-bit serves form the atomic layer of the serve arsenal. Typically delivered short, they restrict the receiver's attacking options and control the transition into I_2.

- **01 (Push)**: Controlled backspin, prioritising stability, depth control and placement.
- **10 (Flick)**: Light, surprising topspin used sparingly to vary tempo and disrupt receiver contact at I_2.
- **11 (Punch)**: Flat, fast contact with complex, residual spin (after the punching action), compresses the receiver's reaction window and forces a rushed return.
- **00 (Graze)**: Deceptive soft float or no-spin contact. Maximum deception, achieved through brushing with negligible normal force ($F_h \approx 0$), producing outcomes dominated by uncertainty.

9.2.2 Intermediate Serves (3-Bit)

Intermediate serves apply immediate pressure or pre-configure the opening rally pattern, often with a designed third-ball attack (I_3) in mind. Table 9.1 summarise the expanded serve space encompassing both basic and intermediate actions.

The 3-bit serve space organises naturally into three tactical clusters:

Table 9.1 Basic and intermediate serves in TTNS

Code	Visual	Serve	Description
01	$\partial 1$	s–Push	Controlled backspin
10	1σ	s–Flick	Light topspin surprise; initiative-seeking
11	11	s–Punch	Flat, fast contact; compresses reaction window
00	$\chi\partial$	s–Graze	Float/no-spin; inducing pop-ups
111	111	s–Smash	Fast, deep pressure serve
110	11σ	s–Drive	Fast, deep, light topspin; sets up third-ball attack (I_3)
100	$1\sigma\sigma$	s–Topspin	Invite a block return for a controlled third-ball loop (I_3)
011	$\partial 11$	s–Poke	Fast backspin cut; force a long, high return suitable for a smash (I_3)
010	$\theta 1\theta$	s–Swipe (or *Lollipop*)	Pure sidespin; disrupts timing and footwork
	$\theta 1\sigma$	s–Topspin Fade	Side-topspin; difficult to time
	$\theta 1\partial$	s–Backspin Fade	Side-backspin; difficult to control
001	$\partial\partial 1$	s–Chop	Heavy backspin forcing a vulnerable high lift (I_2)
000	$\chi\chi\chi$	s–Float	Dead ball inducing a mis-hit or over-hit through absence of spin input

- **Power/topspin dominant**: Smash (111), Drive (110), Topspin (100).
- **Backspin-dominated**: Poke (011), Chop (001), Float (000).
- **Sidespin-dominated**: Swipe (010) and its derivatives, Topspin Fade and Backspin Fade.

The Block (101) is excluded from the serve space, as it is inherently reactive and presupposes an incoming ball.

To distinguish serves from rally strokes, serve actions are prefixed with s–. For example, *s–Push* (01) denotes a serve, whereas *Push* (01) denotes a stroke. In practice, however, the prefix is often unnecessary, as context typically disambiguates the code.

9.2.3 Advanced Serves (4-Bit)

Because service regulations limit maximum speed, effectiveness depends increasingly on spin variation, contact geometry and surprise. TTNS provides a precise language for analysing such techniques without recourse to descriptive folklore.

A key insight is that nearly identical body and arm motions can produce radically different outcomes depending on force magnitude, contact point and brushing direction.

Pendulum Serve (0100)

The pendulum serve is constructed from a Swipe (010) followed by a terminal Brush, yielding the 4-bit code 0100. Variants differ only in the final brushing direction:

- **Sidespin pendulum**: Rθ1$\theta\theta$, dominant sidespin.
- **Backspin pendulum**: Rθ1$\theta\partial$, side-backspin.
- **Topspin pendulum**: Rθ1$\theta\sigma$, side-topspin.
- **Disguised pendulum**: ∂1$\chi\chi$, disguised weak backspin or no-spin.

Reverse Pendulum Serve (0100)

A mirrored pendulum motion produces opposite-direction sidespin outcomes:

- Lθ1$\theta\theta$, Lθ1$\theta\partial$, Lθ1$\theta\sigma$, ∂1$\chi\,\chi$.

Tomahawk Serve (0110)

Derived from the Fast Swipe family, Tomahawk serves are forehand-oriented with a geometrically expansive swing path:

- θ11θ: Heavy sidespin.
- θ11∂: Side-backspin.
- θ11σ: Side-topspin.

The full classification of these advanced serve families is summarised in Table 9.2.

Table 9.2 Advanced serves in TTNS

Code	Visual	Serve	Description
s-0100	R$\theta 1\theta\theta$	Pendulum sidespin	Classic pendulum serve; strong lateral curve with dominant sidespin
	R$\theta 1\theta\partial$	Pendulum side-backspin	Lower-side lateral brushing producing heavy side-backspin
	R$\theta 1\theta\sigma$	Pendulum side-topspin	Upper-side lateral brushing producing a topspin-accented pendulum
	$\partial 1\chi\chi$	Pendulum backspin	Lower-side brushing producing disguised backspin
s-0100	L$\theta 1\theta\theta$	Reverse pendulum sidespin	Mirror pendulum motion; outward brush producing opposite-direction sidespin
	L$\theta 1\theta\partial$	Reverse pendulum side-backspin	Reverse pendulum with a dominant backspin component
	L$\theta 1\theta\sigma$	Reverse pendulum side-topspin	Reverse pendulum with a topspin accent
	$\partial 1\chi\chi$	Reverse pendulum backspin	Disguised reverse pendulum backspin
s-0110	$\theta 11\theta$	Tomahawk sidespin	Forehand-oriented heavy sidespin producing a wide-angle serve
	$\theta 11\partial$	Tomahawk side-backspin	Downward fade producing strong side-backspin
	$\theta 11\sigma$	Tomahawk side-topspin	Upward fade producing fast, accelerating side-topspin

9.3 Combinations and Strategic Pruning

The transition from descriptive categories to quantitative notation reveals the true scale of tactical possibility in serving. Combining a conservative set of 22 core serve types (4 basic, 7 intermediate, 11 advanced) with the 13 placement zones defined in Fig. 10.1 generates a theoretical field of 186 distinct serve variations.

$$4 \text{ basic serves} \times 6 \text{ zones} \; + \; (7 + 11) \text{ serves} \times 9 \text{ zones} \; = 186$$

This quantification explains why serving remains one of the most creative domains of the game. It also exposes a competitive symmetry: this combinatorial space is theoretically accessible to *both* players. Novelty therefore offers only transient advantage; any serve that can be conceived can, in principle, be learned, anticipated or replicated.

Strategic pruning

This leads to a critical principle: *strategic pruning*. Attempting to cycle through a large fraction of 186 permutations induces *serving fatigue*, degrades execution fidelity, and yields diminishing returns once patterns are recognised. Effective serving depends not on exhausting the combinatorial space but on perfecting a compact, deceptive and resilient core repertoire, a personalised serve system constructed from a high-fidelity subset of TTNS codes (e.g., s–0100, s–011), executed within a stable internal model. These core serves must remain effective even when anticipated, capable of competing across diverse opponent skill sets and adaptations.

TTNS as a tactical analysis and encoding tool

The guiding principle from Sect. 4.4.2, 'do what troubles the opponent' is inherently information dependent. Identifying an opponent's vulnerabilities requires sustained observation and rapid in-match synthesis. TTNS functions here as a tactical analysis and encoding tool: recording serves and returns in binary notation renders emerging patterns immediately visible, enabling evidence-based tactical adjustment rather than intuitive guesswork during the rally's opening phase.

9.4 Encoding the Start of a Rally

Consider a rally deliberately designed to produce a third-ball attack. A player serves a pendulum side-backspin to the receiver's crossover point.

Traditional description

Player A served a heavy pendulum backspin wide to the elbow, received a high return and smashed the third ball.

Clear, intuitive and immediately readable.

The TTNS record – a data-driven narrative:

1. **$\mathbf{I_1}$: the designed serve**

$$\text{Serve} = 0100 \mid \text{R}\theta 1\theta_{\partial} \mid \text{Crossover} \mid (\varepsilon_1 = 4, \varepsilon_0 = 11)$$

Interpretation: A high-spin, disruption-oriented serve ($\varepsilon_0 = 11$) intended to constrain the I_2 return. Code 0100 (Rθ1$\theta\partial$) denotes a side-brush-dominant action with deceptive backspin, consistent with a Tactic0 opening.

2. **$\mathbf{I_2}$: the engineered return**

$$\text{Receive} = 01 \mid (\varepsilon_1 = 1, \varepsilon_0 = 2) \mid \text{Height: High} \mid \text{Depth: Long}$$

Interpretation: The serve successfully shapes the resulting rally state. The receiver produces a high, long push, mechanically weak and geometrically exposed. The combined strength, $\varepsilon = \varepsilon_I + \varepsilon_0 = 3$ confirms tactical vulnerability, precisely the target condition for a third-ball attack.

3. **$\mathbf{I_3}$: the executed attack (third ball)**

$$\text{Shot } I_3 = 1110 \mid (\varepsilon_1 = 14, \varepsilon_0 = 1) \mid \text{Outcome: Win}$$

Interpretation: The server converts the engineered asymmetry with a terminal Tactic 1 action. The Whip (1110) delivers dominant translational impulse ($\varepsilon_1 = 14$) with minimal spin reliance, resolving the rally within the same interaction cycle.

Complete TTNS sequence:

$$[I_1\text{: s}-0100,\ R\theta 1\theta_\partial,\ \text{Crossover},\ (4,\ 11)] \rightarrow [I_2\text{: }01,\ (1, 2),\ \text{High},\ \text{long}] \rightarrow [I_3\text{: }1110,\ (14, 1),\ \text{Win}]$$

This compact line encodes the full tactical architecture: designed intent, induced response and decisive resolution. By compressing mechanics, spin, placement and intent into symbolic form, TTNS renders play legible, comparable and systematically improvable.

- **For coaches**: Targeted correction, e.g., the limitation may be insufficient ε_1 in the 011 serve, not inappropriate serve selection.
- **For players**: Clarified personal leverage, e.g., a pendulum with $\varepsilon_0 = 11$ reliably destabilises opponents whose ε_{T0} degrades under crossover pressure.
- **For analysts**: Actionable data. What is the win probability when opening with s–0100 ($R\theta 1\theta\partial$) to the crossover under $\varepsilon_0 \geq 9$?

In the opening exchange of a rally, TTNS provides a mechanism for designing what follows.

The traditional account tells the story. TTNS reveals the structure.

Chapter 10
Variation in Play

A rally evolves through the systematic manipulation of fundamental match parameters. Variation is not random; it is the deliberate injection of informational entropy, designed to degrade the opponent's predictive model. TTNS provides the structure to quantify this entropy across every dimension of play.

10.1 Variation in Placement

Placement is the most direct form of variation. Dividing the table into thirteen weighted zones (Fig. 10.1) enables quantitative evaluation of spatial disruption during a rally.

The placement variation (δ_z) is defined as the absolute difference between zone indices.

- δz (Small Trapezoid $\rightarrow$ Wing, parallel) $= |3 - 6| = 3$
- δz (Deep $\rightarrow$ Small Trapezoid, diagonal) $= |7 - (-3)| = 10$

Large δ_z values impose greater demands on footwork, balance and recovery.

Zone weighting *contextualises* δ_z. A displacement towards an opponent's strong wing carries a lower effective disruption than the same geometric shift towards a known weakness. Thus, δ_z measures not just distance but targeted spatial discomfort.

X. Ji, *Spin, Speed and Strategy*, Mathematical Engineering,
https://doi.org/10.1007/978-3-032-25518-1_10

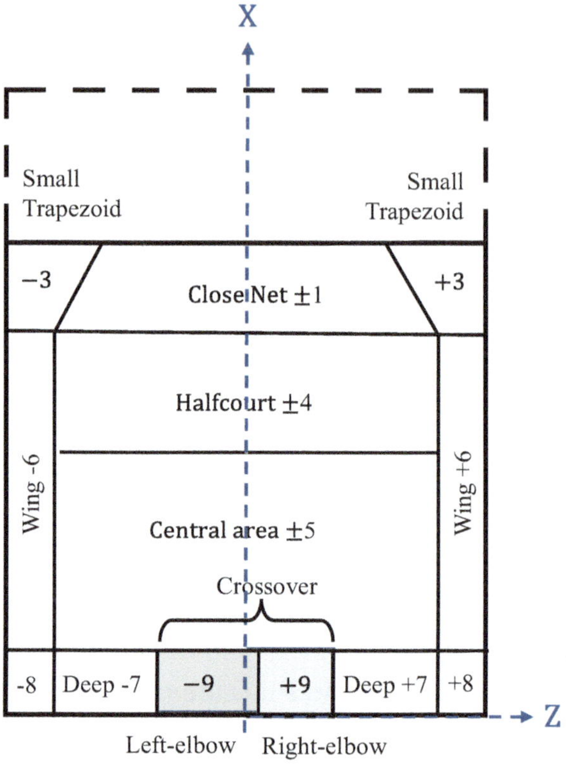

Fig. 10.1 A weighted tactical grid of the half-table assigning numerical values (e.g. ±1 to ±9) to placement zones to quantify shot difficulty

10.2 Variation in Interception Timing

Where Sect. 3.4.1 establishes *interception timing* (τ) as a qualitative tactical choice, this section formalises timing as a quantifiable control variable. By assigning ordinal values to interception phases, timing variation becomes measurable and strategically exploitable.

Ordinal scale of interception timing

- **Rising interception**: $\tau_{\text{rising}} = 7$. Aggressively compresses the opponent's reaction window (t_w).
- **Peak interception**: $\tau_{\text{peak}} = 3$. Maximises stability, controllability and margin.
- **Falling interception**: $\tau_{\text{falling}} = -3, -5, -7$ corresponding to close-to-table, mid-distance and far-distance positions. Increasingly negative values expand the temporal margin available for recovery, intention formation and tactical disruption.

Temporal variation (δ_τ) is defined as the absolute difference between two ordinal values. For example:

- **Peak → Rising**: $|3 - 7| = 4$, a meaningful acceleration of tempo
- **Rising → Falling**: $|7 - (-3)| = 10$, a radical deceleration, often perceived as a sudden loss of rhythm.

Manipulating δ_τ is the essence of temporal disruption. By denying the opponent a stable interception reference, a player destabilises anticipation: footwork, timing and stroke preparation. Timing variation therefore operates as a first-class tactical weapon, equivalent in importance to speed (1), spin (0) and placement (z).

Temporal quantities distinguished

'Time' encompasses several distinct concepts:

- **Timing (τ)**: the instant of ball interception relative to the bounce trajectory.
- **Tempo**: the speed at which successive interceptions occur within a rally.
- **Rhythm (r)**: the perceived regularity of this temporal sequence.
- **Reaction window (t_w)**: the interval between perceiving the opponent's stroke and initiating a response.

These quantities operate on different scales but collectively define the temporal structure of play.

10.3 Variation in Speed and Spin

As established in Sect. 6.5, the extended quantification system (E15) maps DQ4 codes onto a fifteen-level strength scale. The parameters ε_1 (speed strength) and ε_0 (spin strength with sign) therefore represent the quantified intensity of a stroke. These values allow successive shots to be compared in a rally.

Defining variation

Let ε_I denote variation in speed and ε_0 variation in spin between two shots. Their absolute differences quantify the change in the ball's intensity signature.

Example: Punch → Push

- Punch (11): $\varepsilon_I = 3$, $\varepsilon_O \approx 0$
- Push (01): $\varepsilon_I = 1$, $\varepsilon_O = -2$
- Speed variation: $\delta_I = |\varepsilon_I(11) - \varepsilon_I(01)| = |3 - 1| = 2$
- Spin variation: $\delta_O = |\varepsilon_O(11 - \text{ə}1)| = |0 - (-2)| = 2$
- Combined variation: $\delta = \delta_I + \delta_O = 4$.

Tactical interpretation

High δ_1 and/or δ_0 values are the hallmark of sophisticated rallies. They indicate that a player is actively shifting the ball's speed and spin intensity, forcing the opponent to continuously recalibrate between speed-dominant and spin-dominant responses.

10.4 Variation in Rhythm

If timing (τ) specifies an *instant*, rhythm (r) governs the temporal structure of a rally over multiple strokes. Rhythm is not defined by a single action but by how successive strokes relate in pace, interval and interruption. In table tennis, as in music, rhythm is established through mutual calibration between participants and is subsequently manipulated to destabilise temporal stability.

Early exchanges stabilise tempo through controlled contact. As familiarity develops, variation in spin, speed or placement disrupts this equilibrium and creates initiative. Rhythm is therefore not merely a pattern but a *negotiated structure* within the rally.

Extending the causal chain

The non-playing interval between points extends this temporal system. Players modulate this interval, shortening it to compress the opponent's recovery window, lengthening it to introduce uncertainty or restore composure. These adjustments function as boundary conditions for the upcoming rally, shaping the opponent's anticipatory state. Their effect intensifies under score pressure, making the pre-serve phase a mechanism for entropy management *before* ball contact.

The mechanical causal chain (Sect. 2.9.1)

$$\mathrm{I}_1 \rightarrow \mathrm{TC}_0 \rightarrow \mathrm{F}_1 \rightarrow \mathrm{TC} \rightarrow \mathrm{F}_2 \rightarrow \mathrm{I}_2,$$

is therefore extended to

$$\text{Pre-serve} \rightarrow \mathrm{I}_1 \rightarrow \mathrm{TC}_0 \rightarrow \mathrm{F}_1 \rightarrow \mathrm{TC} \rightarrow \mathrm{F}_2 \rightarrow \mathrm{I}_2,$$

where the *pre-serve* is a non-contact, state-conditioning phase.

Governing rhythm

Rallies are rarely decided by a single action. They are won by whoever establishes the governing rhythm, sustain it under pressure and, crucially, break it without losing internal stability. Players do not just react within a rally; they enter, negotiate and reshape its temporal structure. At the highest level, players impose rhythms aligned with their internal model, forcing opponents to play *out of time*.

TTNS formalises rhythm as a discrete variable, *r*, encoded on a 15-level binary scale (Table 10.1). Four anchor tempos provide perceptual and tactical reference points.

- **r_1(Level 1)-Steady and Methodical**: Low-tempo stabilisation and reset. Entropy is suppressed. Pushes, soft blocks and controlled placement dominate.
- **r_{11}(Level 3) -Brisk and Cooperative**: Mutually sustained rhythm. Exchanges are quick but predictable. Latent asymmetries accumulate beneath apparent stability.

Table 10.1 Rhythm levels in TTNS notation

Notation	Level	Descriptor	Practical meaning in match play
r_1	1	Steady	Slow, deliberate tempo; reset and stabilise
r_{10}	2	Methodical	Slightly increased pace; control oriented
r_{11}	3	Brisk	Exchanges quicken; reaction window (t_w) begins to compress
r_{100}	4	Surging	Noticeable tempo surges; no instability
r_{101}	5	Turbulent	Abrupt accelerations; rhythm uneven
r_{110}	6	Jagged	Sustained but irregular tempo; predictable unpredictability
r_{111}	7	Volatile	*Anchor instability point; rhythm resists settling*
r_{1000}	8	Adaptive chaos	Controlled variation; tempo shifted intentionally
r_{1001}	9	Sustained power	Consistent pressure through steady strength
r_{1010}	10	Structured power	Power applied methodically
r_{1011}	11	Intelligent power	Power with intelligent tempo variation
r_{1100}	12	Dynamic steady	Predictable surging tempo
r_{1101}	13	Dynamic methodical	Structured acceleration patterns
r_{1110}	14	Fluid dynamic	Fast, fluid, highly energetic exchanges
r_{1111}	15	Terminal	Maximum tempo; explosive resolution

- **r_{111}(Level 7)-Volatile and Turbulent**: The critical instability zone. Tempo fluctuates abruptly; spin, pace and placement shift without warning. Favours Tactic0 (induced uncertainty).
- **r_{1111}(Level 15)-Explosive and Terminal**: Maximum intensity. Reaction windows (t_w) collapse; points resolve within one or two exchanges. Favours p10/p11 (immediate resolution).

Between these anchors lie intermediate states (Table 10.1) that describe how tempo is shaped, modulated and transitioned.

Rhythm as a strategic variable: δ_r

Players who thrive in the turbulent mid-range (approximately Levels 5–8) specialise in *orchestrated instability*. Their strength lies not in speed alone but in δ_r, the controlled shifting of rhythm that prevents the opponent from settling.

This is why rhythm is inseparable from entropy management:

- Sustained rhythm $\rightarrow$ suppresses entropy
- Broken rhythm $\rightarrow$ entropy amplified

Tactical dominance is achieved when a player preserves internal stability whilst continuously altering the external pulse imposed on the rally. This full rhythm scale is summarised in Table 10.1.

10.5 Total Variation (δ)

If individual variations are instruments, *total variation (δ)* is the full orchestra. It quantifies the aggregate informational and mechanical load imposed on an opponent per shot transition.

Definition

$$\delta = \delta_z + \delta_\tau + \delta_l + \delta_O + \delta_r$$

- δ_z = variation in placement (zone shift)
- δ_τ = variation in interception timing
- δ_l = variation in speed strength
- δ_O = variation in spin strength
- δ_r = variation in rhythm.

Interpretation of δ

- **Low δ**: Stable, patterned exchanges. The domain of the p1 player imposes an orderly tactical model.
- **Sudden δ spike**: A single large shift (e.g., wide placement or disguised serve), representing a designed disruption event.
- **Sustained high δ**: Consistently high variation across multiple transitions. The signature of the p01 chaos player prevents the rally from ever reaching predictability.

Total variation functions as the unifying metric of TTNS. It operationalises entropy, quantifies the Tactic0 strategy and fingerprints playing styles. It provides a single scalar value that captures the total complexity of tactical disruption.

10.6 TTNS as a Hierarchical Language

TTNS is a hierarchical language in which competitive play is represented as a cascade of state transformations, linking physical interaction to tactical consequence and ultimately to match outcome. Each layer encodes the same process at a different level of abstraction, compressing complexity whilst preserving causal structure. Rather than describing isolated techniques, the framework formalises how actions create mechanical and tactical demands, how those demands shape decisions, and how decisions propagate through rallies (Fig. 10.2).

The hierarchy may be interpreted as a causal progression: physical state defines the system configuration, action construction generates state transitions, tactical state encodes the resulting constraints, and match interaction governs temporal evolution.

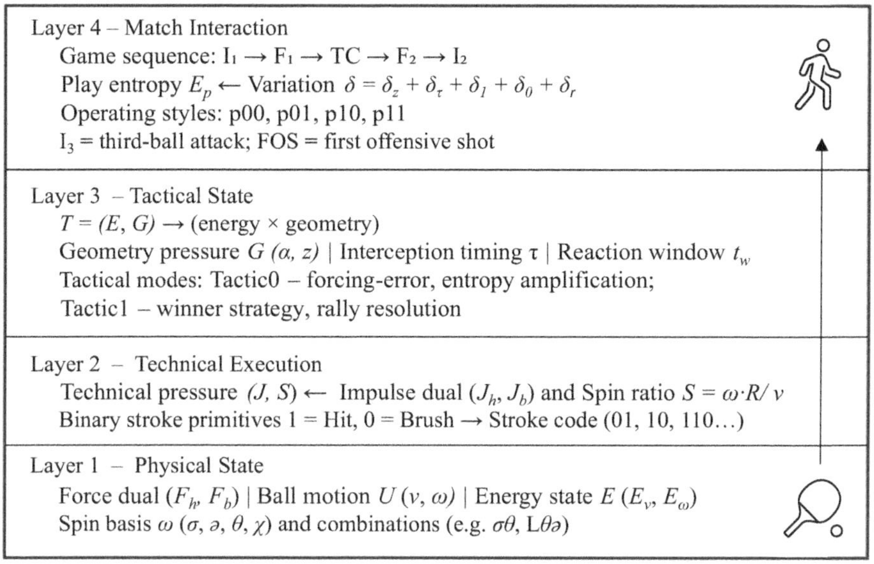

Fig. 10.2 Hierarchical structure of the table tennis notation system (TTNS), linking physical state, action construction, tactical state and match interaction. The layers form a causal progression from mechanical interaction to informational outcome, enabling rallies to be analysed as state transitions within a coupled dynamical system

Layer 1: Physical state

At the foundation lies the physical state of the system, defined by the force dual (F_h, F_b) and the kinematic variables $U = (v, \omega)$, from which the energy components (E_v, E_ω) and spin configuration (σ, ə, θ, χ) are derived. These variables specify the instantaneous condition of the ball.

Layer 2: Action construction

Through player action, this state is transformed via the impulse pair (J_h, J_b), giving rise to structured stroke construction. Binary primitives 1 (Hit) and 0 (Brush) form ordered stroke codes such as 01 (Push), 10 (Flick), 110 (Drive) and 011 (Poke), encoding the sequence of contact mechanics. Stroke realisation is further characterised by the spin ratio

$$S = \omega \cdot R/v,$$

linking mechanical input to resulting motion.

Layer 3: Tactical state

At the tactical level, mechanical outcomes acquire strategic meaning. The state of a rally is defined by

$$T = (E, G),$$

where $E = f\,(v, \omega)$ represents the energetic demand of the return and $G = g\,(\alpha, z)$ represents geometric pressure through placement and trajectory. Rally evolution is governed by variation:

$$T_{n+1} = T_n + \delta,$$

where $\delta = \delta_z + \delta_\tau + \delta_I + \delta_O + \delta_r$. Within this domain, players operate through two complementary modes: Tactic0, which prolongs interaction whilst amplifying entropy, and Tactic1, which resolves the rally through decisive energy, placement or timing pressure. Their balance defines the operating style along the p0–p1 continuum.

Layer 4: Match interaction

At the level of match interaction, rallies unfold as observable sequences

$$\mathrm{I}_1 \rightarrow \mathrm{F}_1 \rightarrow \mathrm{TC} \rightarrow \mathrm{F}_2 \rightarrow \mathrm{I}_2,$$

within which tactical states evolve and informational complexity accumulates as play entropy E_p. Key transitions, such as the first offensive shot (FOS) and third-ball attack (I_3), mark shifts in initiative within this sequence.

The causal chain

Taken together, the hierarchy defines a causal chain linking physical interaction to informational outcome:

$$(F_h, F_b) \rightarrow (J_h, J_b) \rightarrow T(E, G) \xrightarrow{\delta} E_p \quad (10.1)$$

This expression is not symbolic convenience; it is the governing causal structure of the system.

Two governing principles

Two principles summarise the framework:

- **Tactical state**: The state of a rally is defined by $T = (E, G)$, where energy determines physical demand and geometry determines the biomechanical constraints under which that demand must be met.
- **Rally evolution**: Rallies evolve through variation. Each shot modifies the tactical state by an increment δ, and through successive transitions this variation shapes both informational complexity and outcome.

Conclusion

Through this hierarchy, the apparent complexity of table tennis becomes intelligible. The sport can be understood as a coupled dynamical system in which physical state, technical action and tactical demands evolve together. TTNS therefore functions as a generative framework connecting microscopic mechanics with macroscopic competitive behaviour.

10.7 Equipment, Style and Entropy in the Layers of TTNS

At the highest layer of the TTNS hierarchy, the match-interaction layer, equipment extends beyond ball–bat interaction to shape playing styles and influence the entropy landscape of rallies. Equipment selection therefore operates not only at the level of action construction but at the level of higher-order interaction dynamics.

Within TTNS, styles cluster along the p0–p1 continuum.

- **p1-dominant systems** emphasise speed compression and direct resolution as in modern inverted-rubber topspin play.
- **p0-dominant systems** emphasise spin disruption, variation and entropy generation.

Equipment is therefore *a structural selection of tactical regime.* Blade stiffness, sponge compliance, friction coefficient and surface geometry collectively determine the region of the p0–p1 spectrum that a player can occupy effectively.

At the interaction level, rallies map onto four style-entropy states:

- **p11: speed versus speed**, producing high-tempo attacking exchanges.
- **p00**: **disruption versus disruption**, producing control-oriented play defined by geometric pressure $G(\alpha, z)$ and sustained variation $\delta = (\delta_z, \delta_\tau, \delta_I, \delta_O, \delta_r)$.
- **p10: speed versus disruption**, where offensive dominance is mediated through directional variation.
- **p01: disruption versus speed**, where defensive variation disrupts tempo and initiative.

These interaction states explain why equipment mismatches often produce disproportionate outcomes. A player optimised for a stable p11 regime may encounter structural difficulty when forced into a p00 or p01 interaction generated by long-pimple or anti-spin surfaces. The opponent is transforming the informational structure of the rally by modifying spin, disrupting speed stability and altering the effective tactical state $T(E, G)$.

Pimpled rubbers illustrate this mechanism clearly. Long pips, short pips and no-sponge pips differ in sponge thickness, frictional behaviour and spin response, but they share a common strategic effect: they tend to amplify Tactic0, the error-forcing mandate. Long pips are especially disruptive against fast topspin because they can return the ball with altered or reversed spin, transforming an intended Tactic1 attack into a low, error-forcing ball. Short pips and no-sponge pips operate through different mechanisms, including reduced dwell time, flatter rebound, skidding trajectory and timing disruption, but they similarly reduce the opponent's ability to continue a clean topspin, drive or smash sequence.

The historical development of table tennis (Sect. 2.12.5) reflects this coupling. Sponge rubber (1950s) expanded the p1 domain by enabling high spin-speed combinations. Long pimples and anti-spin surfaces represent a complementary migration towards p0, exploiting the very spin dependence introduced by inverted rubbers.

Consequently, equipment selection defines the entropy landscape of a match. Some pairings produce stable, symmetric exchanges, whereas others introduce structural asymmetries that destabilise the rally from the outset. In extreme cases, outcomes are strongly conditioned before extended play develops, not through disparity in skill alone, but through incompatibility of operating regimes within the TTNS style space.

Chapter 11
Match Analysis Via Layered Decomposition

The preceding chapters develop TTNS as a language for understanding, training and representing table tennis. Match analysis is one natural application of that language, but not the only one, nor necessarily the most important. It is included here because it demonstrates the framework in action: layered decomposition reveals the structure of rallies and tactical transitions in ways that other analytical approaches cannot. The goal is illustration, not exhaustiveness.

The analysis examines the men's singles quarterfinal between **Fan Zhendong** and **Tomokazu Harimoto** at the Paris 2024 Olympic Games. The match unfolded under maximal competitive pressure: Fan entered as the sole remaining Chinese contender, whilst Harimoto deployed a relentlessly high-pace, close-to-table attacking system. Despite extreme early score asymmetries (2:11, 11:4), the contest extended to a seven-game decider, offering a rich dataset for multi-layered beyond point-by-point fluctuations.

This layered approach can be interpreted through the tactical state framework, in which rally conditions are represented as $T = (E, G)$. Throughout the match, both players attempted to reshape the tactical state by altering the rally's energetic demands $E\ (v, \omega)$ and the geometric pressure imposed on the opponent $G\ (\alpha, z)$.

Layer 1: Mechanical Interaction (Force–Impulse Structure)

In the opening games, Harimoto established dominance through repeated Brush-dominant actions (ε_0 range: Graze 00 $\rightarrow$ Skim 000) including serves, early timing ($\tau_{\text{rising}} = 7$), compact blocking (101) and a close-to-table stance. These contacts produced spin-loaded and time-compressed impulses that minimised the reaction window (t_w).

Fan initially responded with power-based strokes (i.e., alternating between Whip 1110 and Topspin 1100), attempting to counter rotational impulses with a p11 or p10 strategy. The result was *impulse mismatch* and rising unforced errors.

From Game 3 onward, Fan recalibrated. He reintroduced passive impulse management, using deeper pushes, sway touches and fast soft blocks (i.e., switching among

X. Ji, *Spin, Speed and Strategy*, Mathematical Engineering,
https://doi.org/10.1007/978-3-032-25518-1_11

Poke 011, Sway 00 and Block 101), often directed to the crossover point. These actions embedded p01 logic within Fan's play, restoring mechanical stability and interrupting Harimoto's preferred contact rhythm.

Layer 2: Energy and Entropy

Harimoto's early scoring phases were associated with high-energy rally conditions: short exchanges, rapid acceleration and decisive outcomes within one or two contacts. Although visually frenetic, this pattern is structurally consistent with a system optimised for short-burst output.

As the match progressed, Fan deliberately altered the energetic structure. Through heavier topspin, higher net clearance, deeper placement and extended exchanges, he shifted rallies into a high-δ_z, high-entropy regime. Under these conditions, Harimoto's fast-output system exhibited declining consistency, whilst Fan's stroke quality (J, S) preserved execution fidelity.

In tactical state terms, Fan progressively displaced the rallies from Harimoto's preferred region of $T(E, G)$ towards configurations where energetic sustainability and geometric variation favoured Fan's internal model. The advantage arose not from a single decisive attack but from sustained energetic variation and consistent geometric pressure.

Layer 3: Action Vocabulary (TTNS Compression)

Action compression is the operational consequence of internalising high-reliability stroke subsets and pruning excessive variation. Under pressure, a player contracts the action space, reducing executional entropy whilst preserving functional clarity. The effective vocabulary becomes a tightly bounded, high-fidelity subset of the TTNS space.

As pressure intensified, Fan converged further on this compressed repertoire: short soft serves, controlled placement and Brush-dominant forehand and backhand loops. Variability decreased; decisiveness increased.

Harimoto, by contrast, experienced *action expansion*. Disrupted timing prevented effective pruning, forcing reactive adjustments and rushed Hit-dominant responses outside his optimal regime. His action set became less selective and more diffuse, increasing internal entropy and degrading stroke quality.

The contrast reveals a fundamental asymmetry: the player who sustains action compression maintains a stable, low-entropy internal model, whilst the opponent is driven into expansion, where decision-making and execution become progressively less reliable.

Layer 4: Tactical Mode (Tactic0 vs Tactic1)

Harimoto's early advantage was secured primarily through Tactic0: forcing errors via speed and compressed reaction windows.

As Fan regained control, rallies increasingly resolved through Tactic1 sequences. Points concluded through winners rather than opponent errors. This transition reduced Fan's dependence on volatility and shifted point resolution into a predictable, favourable regime.

Layer 5: Temporal Structure (Rhythm and Timing Control)

Harimoto's system is optimised for iso-rhythmic exchanges at extreme tempo. Fan systematically fractured this structure by varying ball flight, depth and interception timing, disrupting rhythmic predictability and imposing repeated recalibration demands.

In the decisive stages, *temporal misalignment* rather than isolated execution became the dominant determinant of outcomes.

Synthesis: What Actually Decided the Match

This match is best understood not as a momentum-driven comeback but as a controlled systemic transition. Fan progressively reconfigured the mechanical, energetic, tactical and temporal conditions of play until Harimoto's high-pace system could no longer operate within its optimal parameters.

From the perspective of $T(E, G)$, Fan gradually altered both the energetic intensity and the geometry of play, shifting the interaction space away from Harimoto's preferred high-frequency regime.

Victory was secured from a precise asymmetry:

- **Internally**: Fan protected execution by compressing his action set and insisting on a stable model under pressure.
- **Externally**: He imposed variability, mechanical, energetic and temporal, at moments that maximised opponent disruption.

Fan did not win by overpowering Harimoto at his peak pace. He won by deciding which paces would matter.

At the highest level, matches are not decided by isolated technical brilliance but by the capacity to engineer the interaction space itself. Fan's win exemplifies this capacity at its highest level: remaining internally invariant whilst forcing the opponent to adapt continuously.

Methodological Universality

Although the tactical outcomes reflect these two athletes' distinctive profiles, the analytical structure remains general. By identifying the dominant interaction layer at each phase and tracing how adjustments propagate across the competition, layered decomposition provides a rigorous framework for match analysis and preparation. This method is not a template to be copied but a lens through which high-speed competitive interaction becomes intelligible.

Chapter 12
Interpreting the Game as a System

Players often ask a simple question: *what makes table tennis so compelling?* It is clearly more than a fitness exercise or a pastime. Improvement remains possible after many years, yet progress is neither linear nor guaranteed. Mastery appears attainable and perpetually incomplete. This tension is central to its structure.

Having analysed individual mechanisms and tactical elements in earlier chapters, this chapter interprets table tennis as a *structured dynamical system*. The preceding chapters have developed a framework that goes beyond description, constructing a symbolic vocabulary designed to make the invisible structures of the sport visible, measurable and actionable. The aim of this final synthesis is to situate that detailed classification within a broader, systems-level understanding.

12.1 Structural Tensions in Play

The preceding chapters have examined the fundamental physical mechanisms of table tennis: spin, friction, air resistance, rhythm and variation. These interactions are objective, measurable and constraining. Simultaneously, human judgement, anticipation, adaptation and choice, remains irreducible.

Progress does not arise from eliminating these inherent tensions but from learning to operate within them, as players do not transcend physical limits; but negotiate them.

Classical physics has long been sufficient to describe table tennis interactions, yet it has rarely been applied *systematically* to expose their repeatable structures and governing mechanics. By developing TTNS, this book provides a formal representational structure for linking classical mechanics with cognitive execution. When interpreted through this lens, the game reveals itself as a system shaped by persistent opposing tendencies. These recurring tensions and their associated TTNS metrics are summarised in Table 12.1.

X. Ji, *Spin, Speed and Strategy*, Mathematical Engineering,
https://doi.org/10.1007/978-3-032-25518-1_12

Table 12.1 Structural tensions and TTNS metrics

Opposing tendency	Description	TTNS metric and primitive focus
Control versus disruption	Imposing order versus generating disorder	δ (total variation) versus ε (execution/metric strength and precision)
Stability versus entropy	Predictable patterns versus controlled unpredictability	δ_r (rhythm and temporal variation)
Intention versus reaction	Pre-planned action versus adaptive response	τ_{rising} versus τ_{falling} (interception timing dynamics)

Analytical models do not predict outcomes deterministically. They describe empirically observed tendencies. The purpose of this structural perspective is not rigid prediction but orientation: to sharpen qualitative judgement by making the underlying matrix of play visible through the TTNS vocabulary.

12.2 TTNS as a Systemic Control Tool

TTNS functions as a formal representational system for table tennis. It abstracts complex, high-velocity interactions and translates them into quantifiable, comparable structures. This vocabulary operationalises the structural tensions of play in three principal ways:

1. **Quantifying uncertainty (entropy)**

The total variation index (δ), renders the operational disorder of a rally explicit. A so-called 'chaos player' is no longer an analytical mystery; TTNS defines them as an agent whose style is characterised by a consistently high average δ, reflecting deliberate control over engineered variability across spin, velocity, and spatial geometry.

2. **Controlling time (tempo)**

Timing variation (δ_τ) formalises tempo as an explicit asymmetry in decision-making windows. Early timing (τ_{rising}) compresses the opponent's analytical and computational time; late timing (τ_{falling}) expands one's own margin for correction. Through TTNS, tempo is rescued from vague terminology and redefined as a measurable temporal advantage, not speed alone.

3. **Predicting instability**

Small perturbations can dominate macro-outcomes. The TTNS framework exposes how excessive control produces systemic fragility, leaving a player vulnerable to high-δ attacks. Conversely, excessive variation generates instability, increasing the probability of unforced errors. Effective play depends on aligning execution with a small number of governing principles under severe time pressure.

Underlying these capabilities is the discrete structure of TTNS, which provides a rigorous second representation of the game alongside its continuous physical description. Forces, velocities and spin evolve continuously in the physical world, but human actions are organised into recognisable patterns that can be encoded, combined and sequenced. This dual representation allows table tennis to be understood simultaneously as a physical system and as a highly structured language of interaction.

TTNS enables response selection based not on arbitrary, folk-coaching rules but on *systemic necessity*, for example, maximise δ_z and δ_r whilst maintaining a low δ_l to exploit an opponent's quantified limits of variation tolerance.

12.3 Player Development

Beyond match analysis, TTNS offers an integrated architecture for performance development, training design and long-term engagement with the sport.

1. **Adaptability and Uncertainty Tolerance**

Performance improvement is best framed as an increase in *systemic adaptability*: the ability to function across a wide range of interaction states. Within the TTNS paradigm, this involves two complementary capacities:

- **Generate high δ**: The capacity to shift multiple parameters (placement, rhythm, spin) rapidly to disrupt an opponent's internal tactical control.
- **Tolerate high δ**: The capacity to maintain execution fidelity when facing fast, unpredictable or heavily spun incoming states.

The latter is often more decisive. High-intensity training should aim not only at mechanical precision under stable conditions but specifically at raising the player's internal threshold for absorbing disorder without experiencing fidelity loss.

2. **Structural Understanding**

When players grasp the fundamental dialectics of the game and understand the quantitative significance of TTNS metrics, the game begins to appear simpler because its organising principles have finally been formalised.

This understanding enables the player to apply the minimal effective intervention required to alter rally dynamics. Solving complex, high-speed problems with restrained, well-timed actions is a primary source of sustained satisfaction in the sport. To fully understand the game through this framework is to learn to read and write its underlying structure using a unified, scientific vocabulary.

12.4 Future Direction of Match Player

Future elite match play is likely to place still greater tactical weight on the opening exchanges. As serve, receive and the first offensive stroke become more decisive, players will have less time to establish a stable reading of spin, pace and equipment response. This compression increases the value of unfamiliarity in all its forms. Equipment such as long pimples and anti-spin surfaces exploits this directly, altering rebound and timing in ways that cannot be easily decoded within the first three to five balls. The same logic extends to unorthodox techniques, unusual spin profiles and rhythm-breaking patterns: their disruptive effect is concentrated before cognitive adaptation can occur. Future competition may therefore be shaped not by raw physical speed alone, but by who can impose a high-δ tactical state before the opponent's internal model has time to form.

Closing Note

Building on the analyses developed throughout this book, this chapter has synthesised the physical, tactical and systemic principles into an integrated account of the sport: one that is rigorously physical, yet deeply human.

Analysis is necessary, but never sufficient. Video, notation and memory reduce the rally into forms that can be studied, but the live opponent remains richer than any representation. A fresh defeat or difficult match exposes the limits of a player's current internal model and forces adaptation under genuine pressure. TTNS is therefore not a replacement for lived play, but the language for interpreting what lived play reveals. Understanding becomes durable when tested against uncertainty.

Constraint sharpens touch. Hard nets and outdoor tables demand greater clearance and truer spin. What initially frustrates proves most instructive. Much of the insight behind this work has been forged not in training halls or traditional coaching sessions, but in parks, public spaces and real-time competition, most notably in Manchester league matches. The progress described here owes as much to those unpolished environments as to any laboratory or abstract theory.

Even a single game, perhaps lost 8:11, carries meaning beyond the score. Two players exchange dozens of strokes, probing angles, spin and timing, communicating through the ball. In that conversation, each explores for a path to victory whilst striving to remain in the rally itself. Competition and cooperation coexist: the effort to win is inseparable from the shared act of sustaining play.

This book reflects an exhaustive effort to construct a language that describes table tennis with the precision, mathematical respect, and systemic accuracy it deserves, a structure built to honour what the game has given in return: exercise, health, friendship and shared understanding.

GPSR Compliance

The European Union's (EU) General Product Safety Regulation (GPSR) is a set of rules that requires consumer products to be safe and our obligations to ensure this.

If you have any concerns about our products, you can contact us on ProductSafety@springernature.com

In case Publisher is established outside the EU, the EU authorized representative is:

Springer Nature Customer Service Center GmbH
Europaplatz 3
69115 Heidelberg, Germany

Batch number: 10918418

Printed by Printforce, the Netherlands